walkernaths 3.3
TRIGONOMETRY

NCEA Level 3 Internal

Charlotte Walker and Victoria Walker

Walker Maths 3.3 Trigonometry
1st Edition
Charlotte Walker
Victoria Walker

Designer: Cheryl Smith, Macarn Design
Production controller: Alice Kane

Any URLs contained in this publication were checked for currency during the production process. Note, however, that the publisher cannot vouch for the ongoing currency of URLs.

Acknowledgements
Cover photo courtesy of Shutterstock.

We wish to thank the Boards of Trustees of Darfield and Riccarton High Schools for allowing us to use materials and ideas developed while teaching. Our thanks also go to all past and present colleagues, especially Kath Wilson, who have generously shared their experience and ideas.

For product information and technology assistance,
in Australia call **1300 790 853**;
in New Zealand call **0800 449 725**

For permission to use material from this text or product, please email **aust.permissions@cengage.com**

National Library of New Zealand Cataloguing-in-Publication Data
A catalogue record for this book is available from the National Library of New Zealand.

978 0 17 042572 8

Cengage Learning Australia
Level 7, 80 Dorcas Street
South Melbourne, Victoria Australia 3205

Cengage Learning New Zealand
Unit 4B Rosedale Office Park
331 Rosedale Road, Albany, North Shore 0632, NZ

For learning solutions, visit **cengage.co.nz**

Printed in China by 1010 Printing International Limited
6 7 23

CONTENTS

Glossary

Make your own glossary of key terms:

Term	Definition	Picture/Example
Cycle		
Period		
Amplitude		
Frequency		
Radian		
Secant		
Cosecant		
Cotangent		

  ISBN: 9780170425728

Revision of trigonometric functions

You will have learned about three trigonometric functions: **sin x** (sine)
cos x (cosine)
and **tan x** (tangent).

f(x) or y = sin x
In a right-angled triangle, $\sin x = \dfrac{\text{opposite}}{\text{hypotenuse}}$

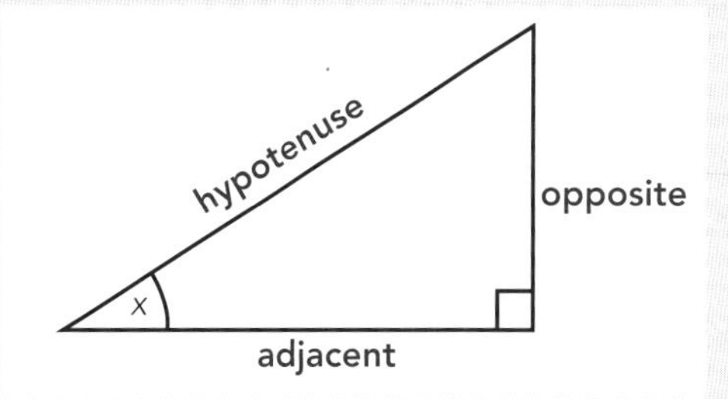

f(x) or y = cos x
In a right-angled triangle, $\cos x = \dfrac{\text{adjacent}}{\text{hypotenuse}}$

f(x) or y = tan x
In a right-angled triangle, $\tan x = \dfrac{\text{opposite}}{\text{adjacent}}$

You should also be familiar with the behaviour of sine and cosine functions for angles less than 0° and greater than 90°.

f(*x*) = sin *x*:

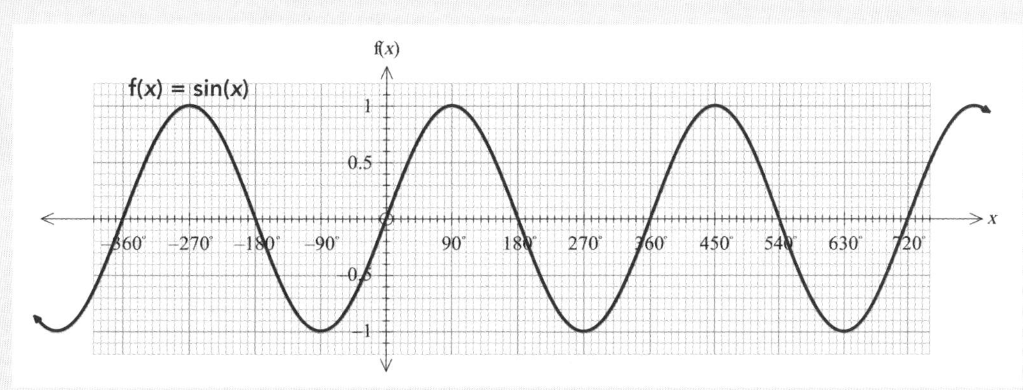

f(*x*) = cos *x*:

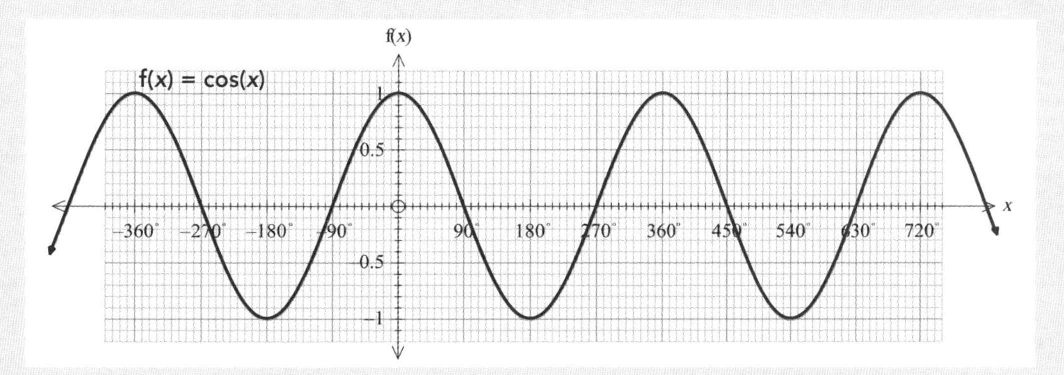

These are known as **periodic** functions.
Instead of measuring angles in degrees, you will be measuring them in **radians**.

ISBN: 9780170425728

Radians

- One radian = the angle formed in a sector with an arc length that is the same as the radius.
- One radian is approximately **57.3°**.

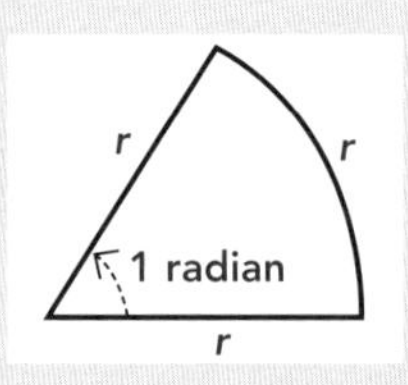

- The length of the radius doesn't make any difference to the angle size:

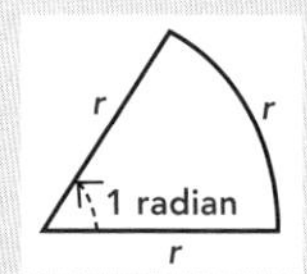

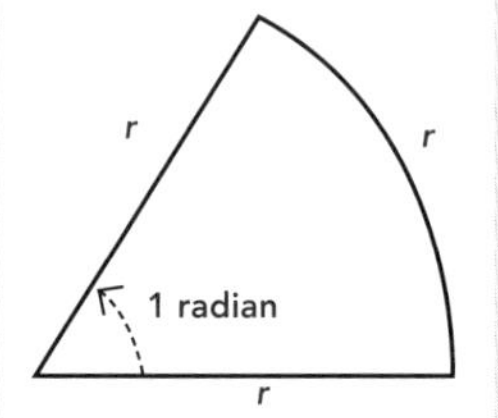

- Make sure you know how to switch your calculator between degrees and radians.

To convert between radians and degrees:

$$\frac{1 \text{ radian}}{360°} = \frac{r}{\text{circumference of the circle}}$$

$$= \frac{r}{2\pi r}$$

$$= \frac{1}{2\pi}$$

$\therefore 2\pi$ radians = 360°

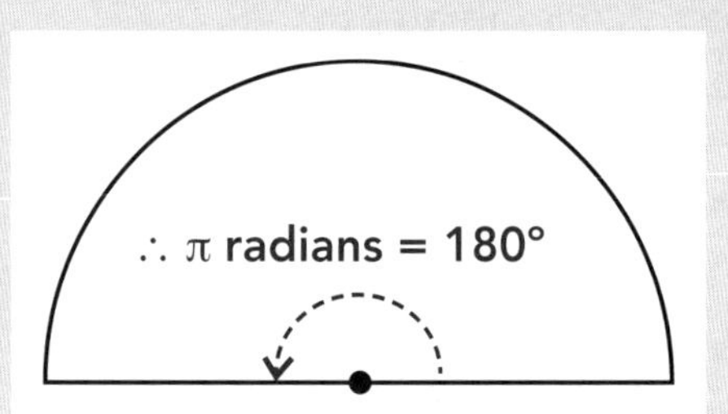

Example: Convert 45° to radians.

180° = π radians

Always use this as your starting point.

$$\therefore 45° = \pi \times \frac{45°}{180°}$$

$$= \pi \times \frac{1}{4}$$

$$= 0.7854 \text{ or } \frac{\pi}{4}$$

It is often convenient to write angles in terms of π.

Example: Convert $\frac{2\pi}{3}$ (radians) to degrees.

Whenever you see an angle which includes π, that angle will be in radians.

π radians = 180°

$$\therefore \frac{2}{3}\pi = \frac{2}{3} \times 180°$$

$$= 120°$$

Example: Convert 7 (radians) to degrees.

π radians = 180°

$$\therefore 7 \text{ radians} = 180° \times \frac{7}{\pi}$$

$$= 401°$$

 ISBN: 9780170425728

Convert the following angles from degrees to radians.

1 $360° =$ ______

2 $90° =$ ______

3 $30° =$ ______

4 $225° =$ ______

5 $540° =$ ______

6 $15° =$ ______

7 $24° =$ ______

8 $99° =$ ______

9 $58° =$ ______

10 $114° =$ ______

Complete the following diagrams, leaving π in your answers.

11

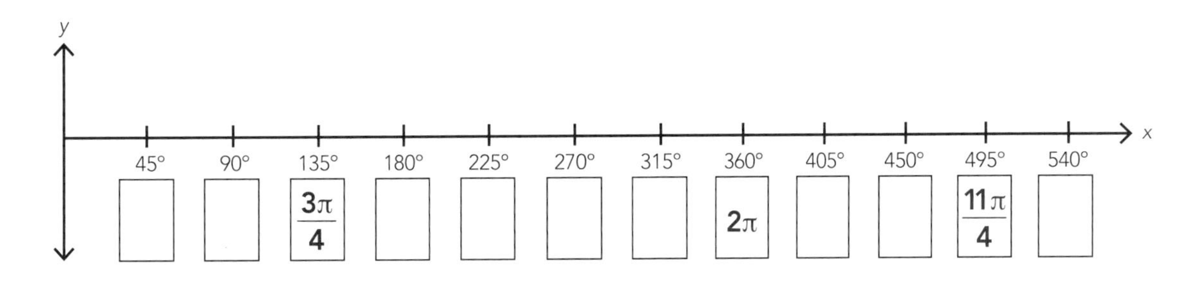

12

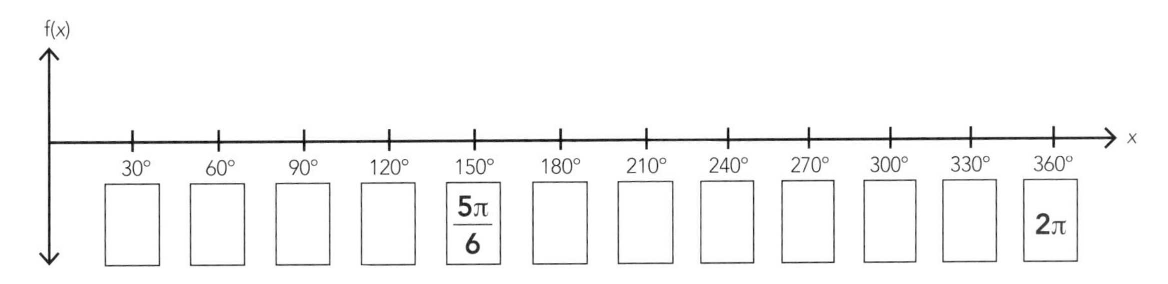

ISBN: 9780170425728

Convert the following angles from radians to degrees.

13 $3\pi =$ __________

14 $\frac{\pi}{2} =$ __________

15 $\frac{\pi}{5} =$ __________

16 $0.3\pi =$ __________

17 $0.3 =$ __________

18 $\frac{7\pi}{18} =$ __________

19 $5 =$ __________

20 $\frac{5\pi}{12} =$ __________

21 $1 =$ __________

22 $\frac{17\pi}{12} =$ __________

Use your calculator to find the following sines, cosines and tangents of angles. Make sure you switch your calculator to **radians** and use **brackets** when there are fractions.

23 $\sin \pi =$ __________

24 $\cos 2\pi =$ __________

25 $\tan \frac{\pi}{2} =$ __________

26 $\sin \frac{3\pi}{2} =$ __________

27 $\cos \frac{\pi}{3} =$ __________

28 $\tan \left(-\frac{\pi}{4}\right) =$ __________

29 $\cos \frac{\pi}{4} =$ __________

30 $\sin \frac{\pi}{4} =$ __________

31 $\cos \frac{5\pi}{12} =$ __________

32 $\sin \frac{4\pi}{3} =$ __________

33 $\cos (-0.7) =$ __________

34 $\sin 3.9 =$ __________

ISBN: 9780170425728

Trigonometric functions

Features associated with periodic functions

Periodic functions

- f(x) = sin *x*, f(x) = cos *x* and f(x) = tan *x* are all **periodic functions**.
- Periodic functions **repeat the same pattern indefinitely**.

Cycle = **one complete pattern**, with no repetition.

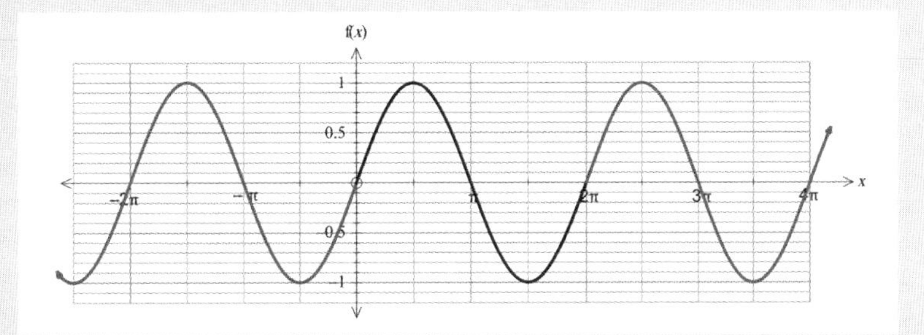

Period = horizontal **distance** required for one complete cycle.

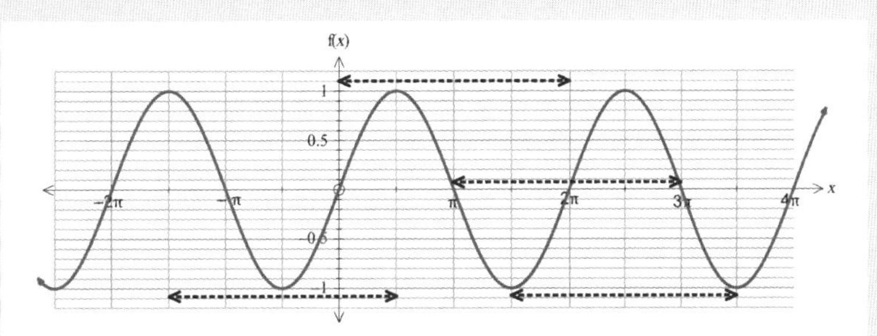

Amplitude = **height** from the centre line to the peak or trough.

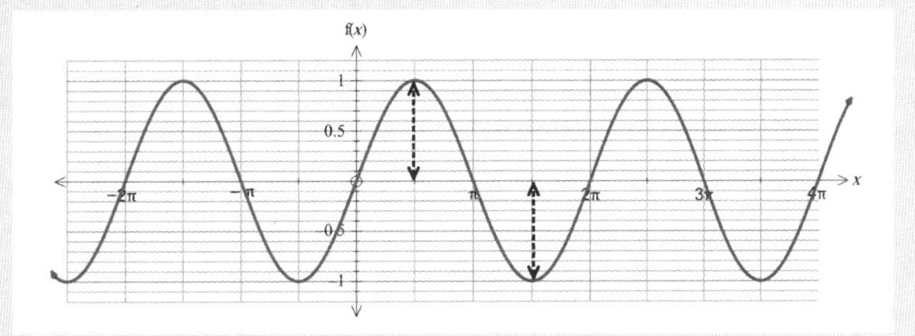

Frequency = the number of cycles in 2π radians = $\dfrac{2\pi}{\text{Period}}$

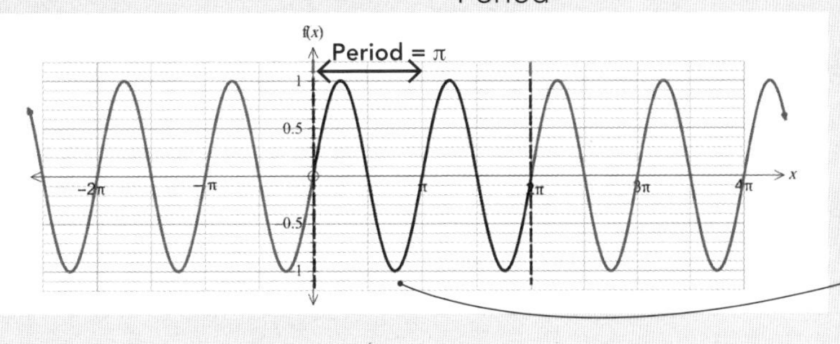

In this case there are exactly two cycles in 2π radians so the frequency is $\dfrac{2\pi}{\pi} = 2$.

ISBN: 9780170425728

Write down the period, frequency and amplitude for the following graphs.

1

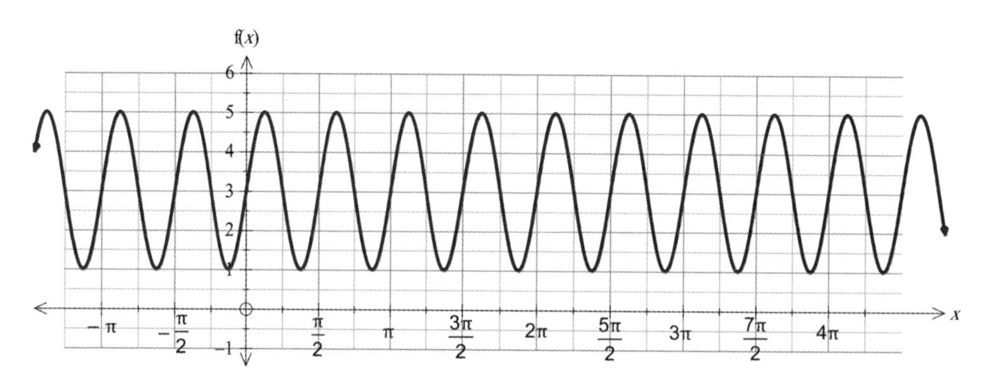

Period = ____________ Frequency = ____________ Amplitude = ____________

2

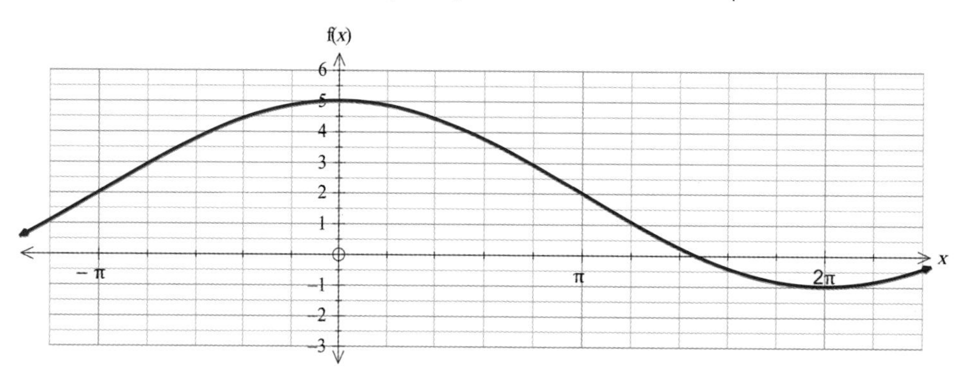

Period = ____________ Frequency = ____________ Amplitude = ____________

3

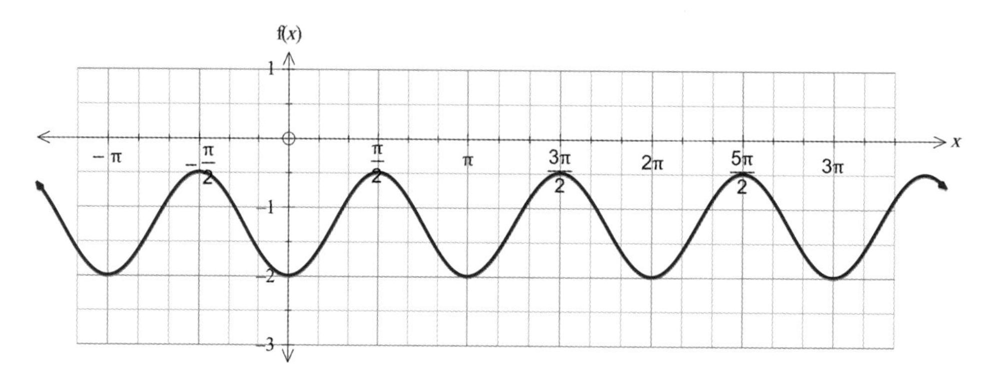

Period = ____________ Frequency = ____________ Amplitude = ____________

4

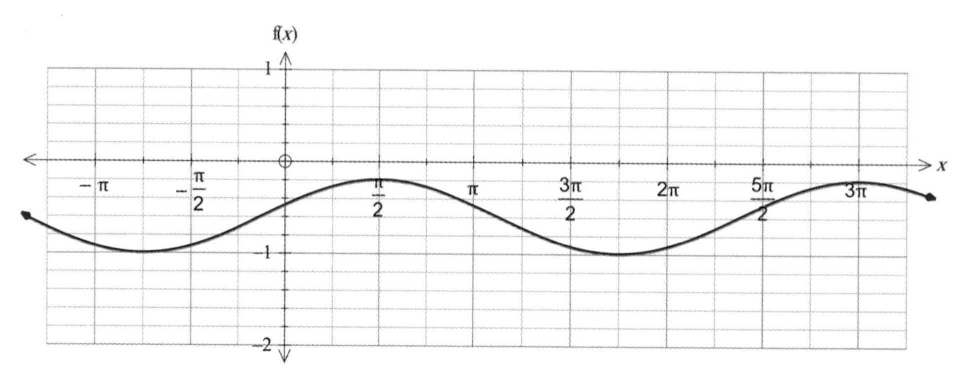

Period = ____________ Frequency = ____________ Amplitude = ____________

ISBN: 9780170425728

The basic trigonometric functions

f(x) = sin x

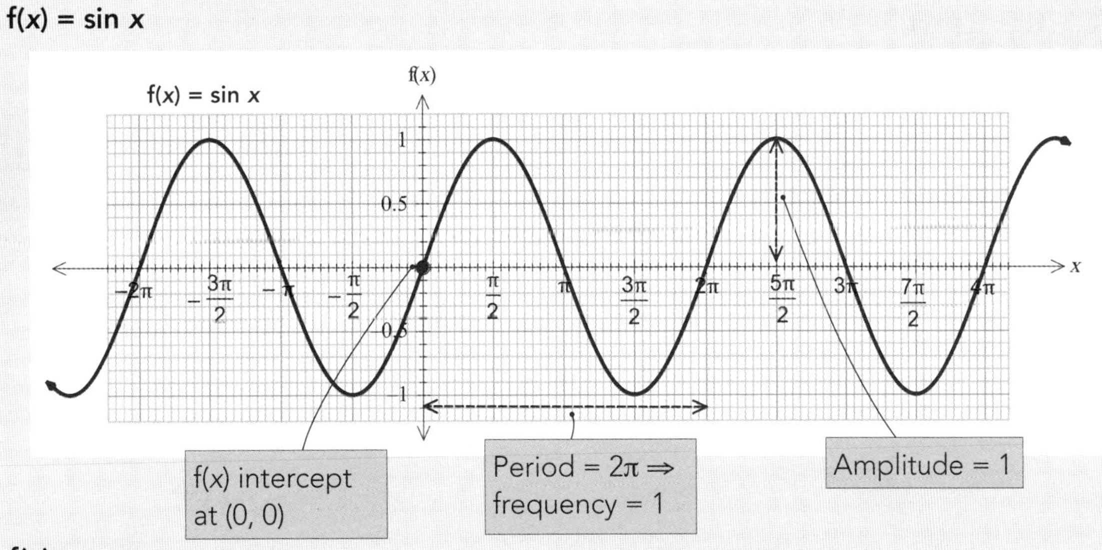

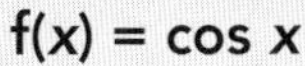

f(x) = cos x

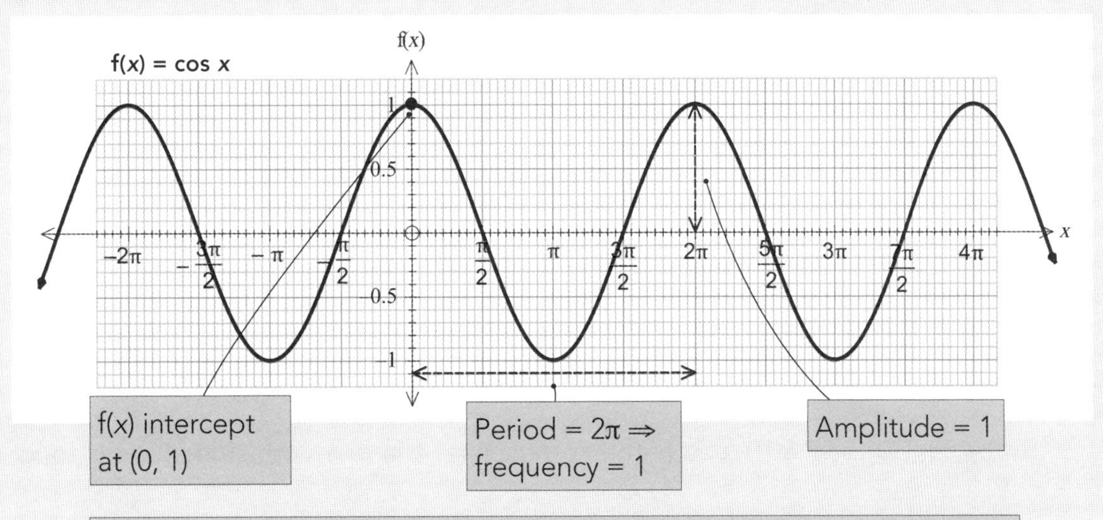

Notice that the sine and cosine graphs are the same shape and they are horizonal shifts of each other: $\sin x = \cos(x - \frac{\pi}{2})$ and $\cos x = \sin(x + \frac{\pi}{2})$.

f(x) = tan x

You will not often meet the graph f(x) = tan x in this standard.

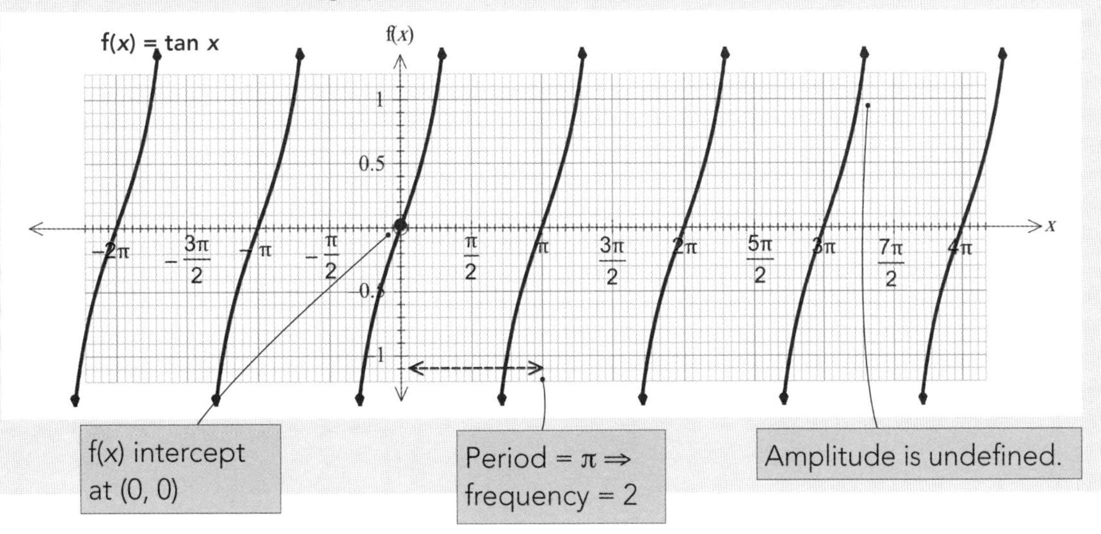

ISBN: 9780170425728

Exact values for trigonometric functions: special triangles

- Trigonometric functions for **most angles do not have exact values**.
- Your calculator gives you these values to many decimal places, and it is usual to **round** these to **4 decimal places**.
  For example: cos 0.6 = 0.825335614… = 0.8253 (4 dp)
- Using **special triangles**, it is possible to write **exact values** for some angles.
- **Exact** values must be able to be written as fractions made up of **whole numbers** and/or **surds** (roots of whole numbers).

For $\frac{\pi}{4}$ (or 45°), use an isosceles right-angled triangle with equal sides 1 unit long:

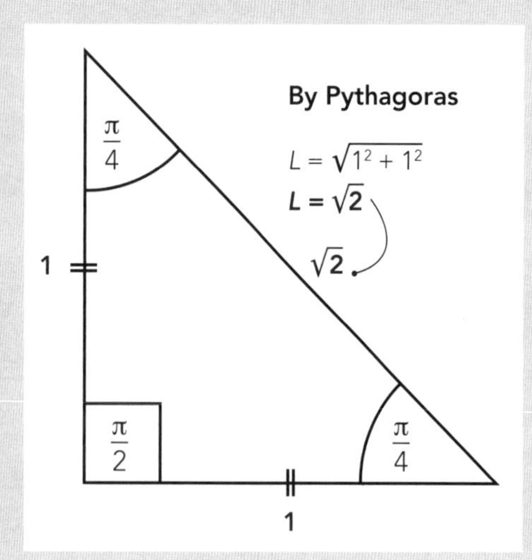

Using this triangle:

$$\sin\frac{\pi}{4} = \frac{1}{\sqrt{2}}$$

$$\cos\frac{\pi}{4} = \frac{1}{\sqrt{2}}$$

$$\tan\frac{\pi}{4} = 1$$

For $\frac{\pi}{6}$ and $\frac{\pi}{3}$ (or 30° and 60°), use half an equilateral triangle with sides 2 units long:

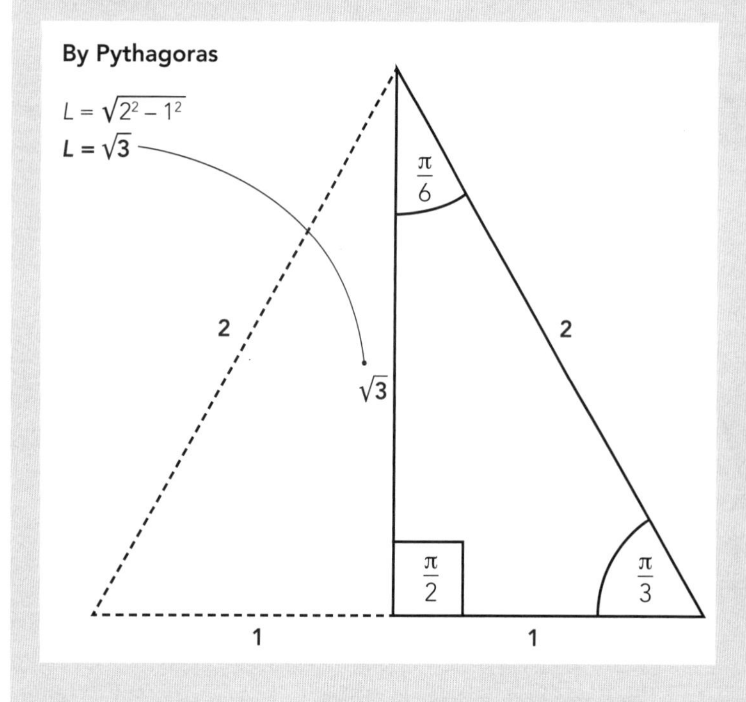

Using this triangle:

$$\sin\frac{\pi}{6} = \frac{1}{2} \qquad \sin\frac{\pi}{3} = \frac{\sqrt{3}}{2}$$

$$\cos\frac{\pi}{6} = \frac{\sqrt{3}}{2} \qquad \cos\frac{\pi}{3} = \frac{1}{2}$$

$$\tan\frac{\pi}{6} = \frac{1}{\sqrt{3}} \qquad \tan\frac{\pi}{3} = \sqrt{3}$$

ISBN: 9780170425728

Another way of viewing trigonometric functions

- It is often convenient to view trigonometric functions in terms of quadrants:

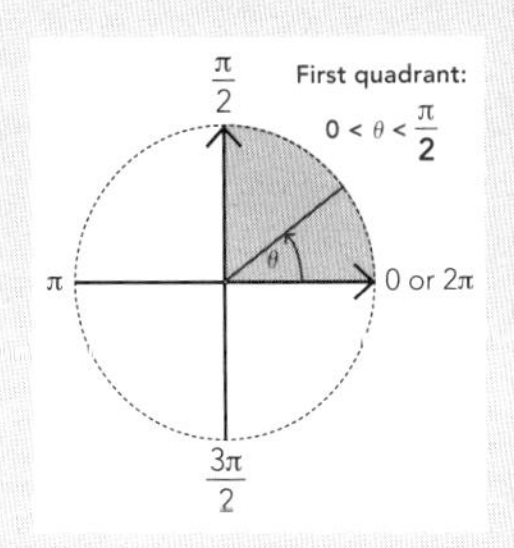

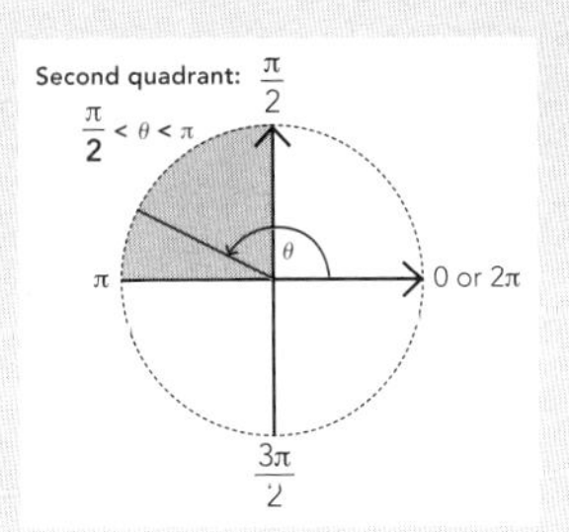

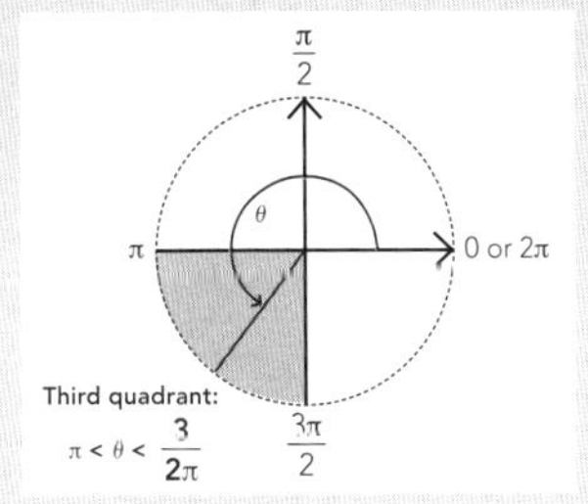

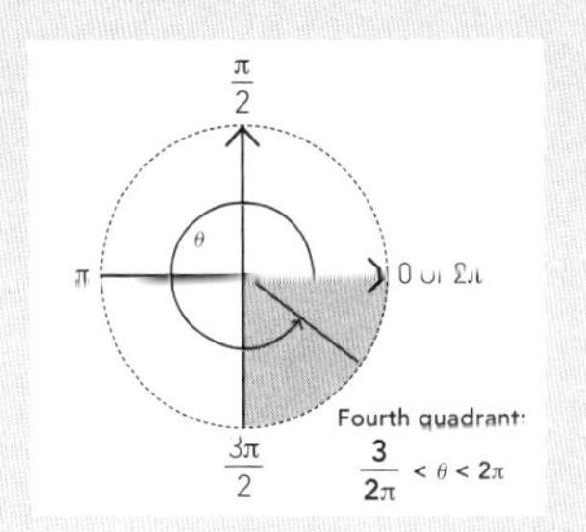

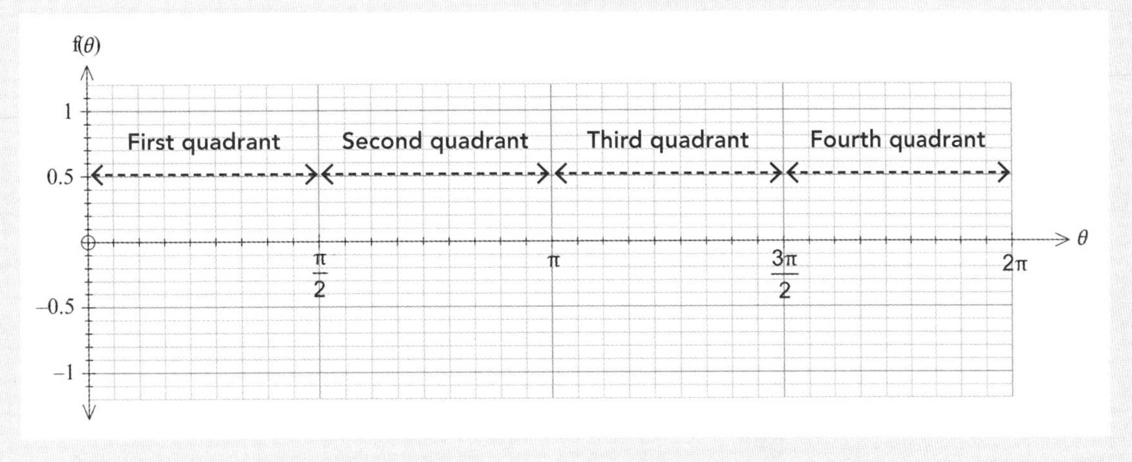

- It is also useful to be aware of the signs of each trigonometric function in each quadrant.

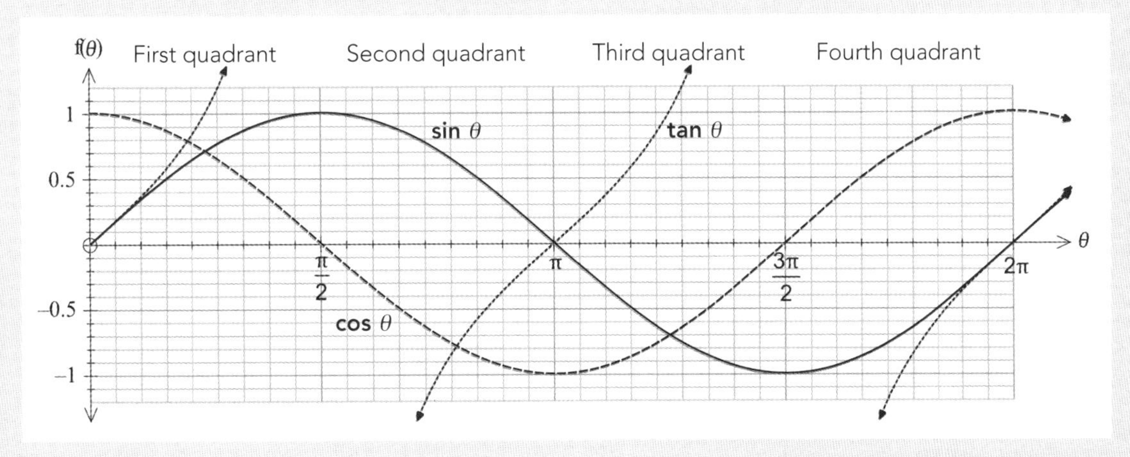

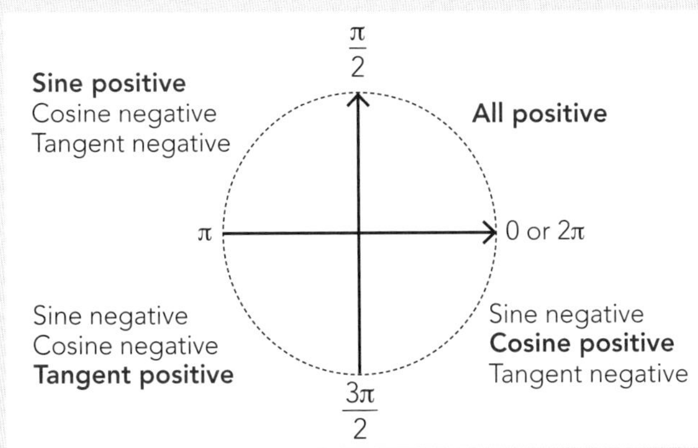

- This approach makes it easy to find other functions when told the value of one.

ISBN: 9780170425728

Examples:

1 a If $\sin\theta = \frac{1}{2}$ and $0 < \theta < 2\pi$, find all possible values for θ.

Sin θ is positive in the first and second quadrants.

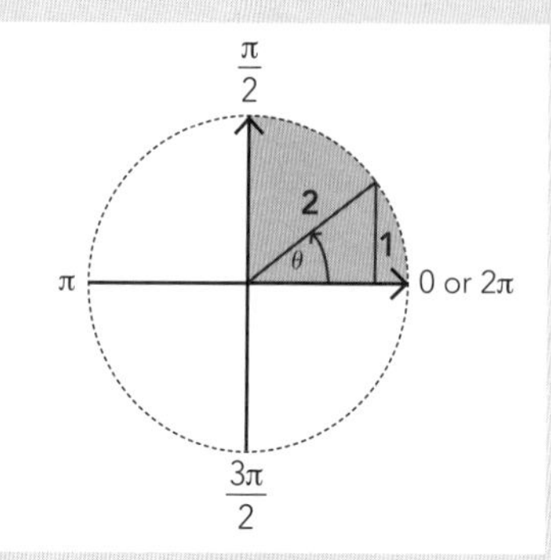

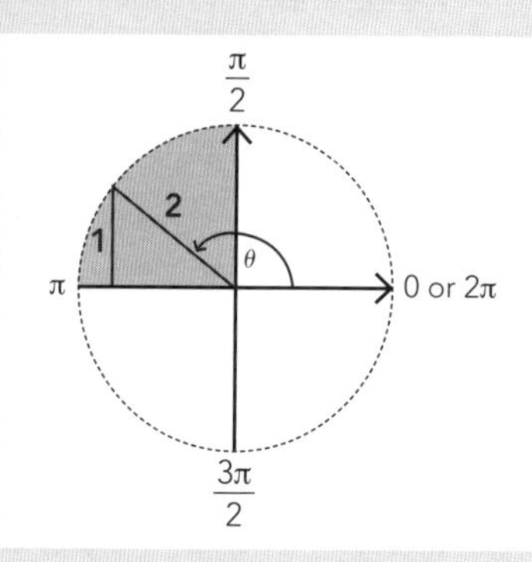

$$\therefore \theta - \frac{\pi}{6} \quad \text{or} \quad \theta = \pi - \frac{\pi}{6} = \frac{5\pi}{6}$$

Note: You could also do this using a graph:

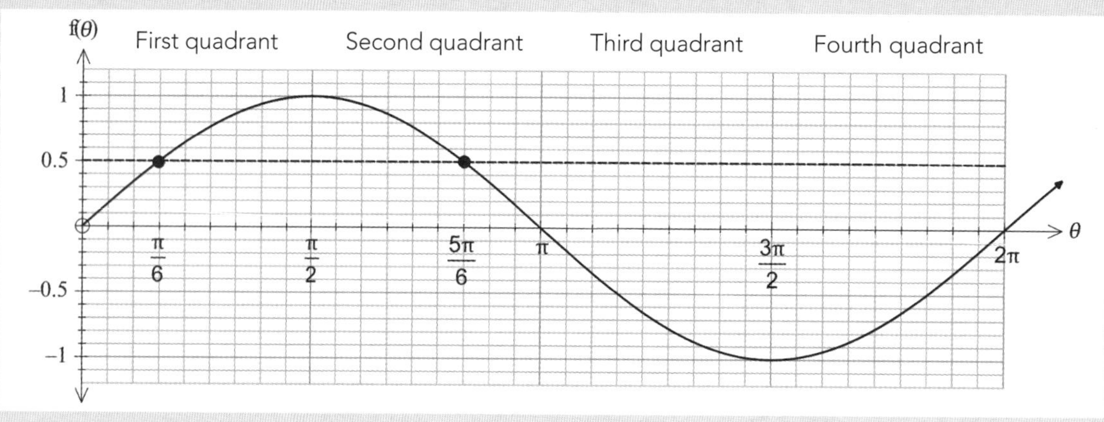

b If $\sin\theta = \frac{1}{2}$ and $0 < \theta < 2\pi$, find the values of $\cos\theta$ and $\tan\theta$.

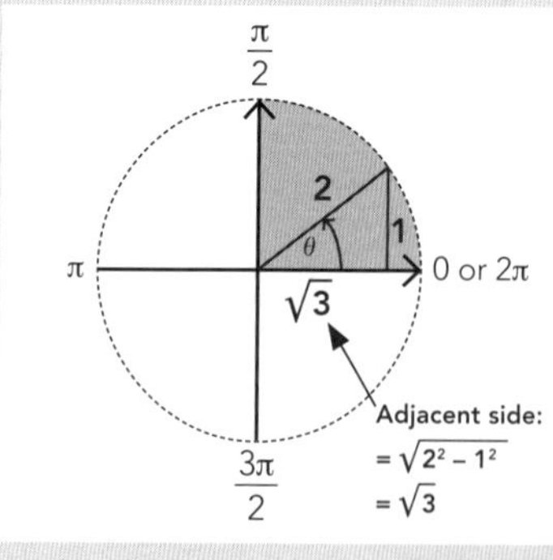

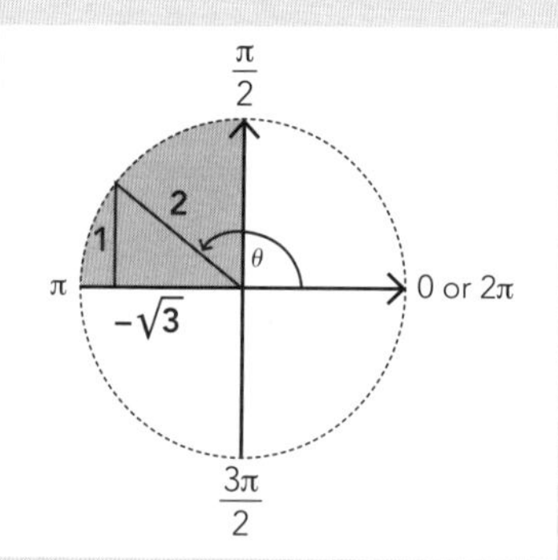

$$\therefore \theta = \frac{\pi}{6} \Rightarrow \cos\theta = \frac{\sqrt{3}}{2} \quad \text{or} \quad \theta = \frac{5\pi}{6} \Rightarrow \cos\theta = \frac{-\sqrt{3}}{2}$$

$$\tan\theta = \frac{1}{\sqrt{3}} \qquad\qquad \tan\theta = \frac{1}{-\sqrt{3}}$$

Remember, when asked for **exact** values, leave root signs in your answer.

Note: The quadrant approach allows you to calculate the values for any trigonometric function, but doing it on a graph does not.

ISBN: 9780170425728

2 If $\tan\theta = -\frac{1}{\sqrt{8}}$ and $0 < \theta < 2\pi$, find the values of $\sin\theta$ and $\cos\theta$.

Tan θ is negative in the second and fourth quadrants.

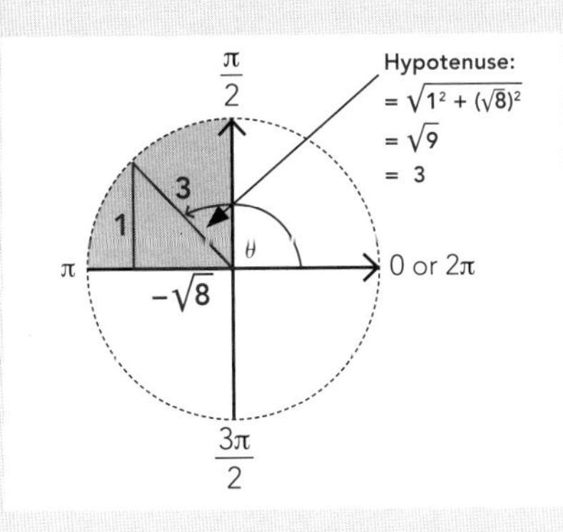

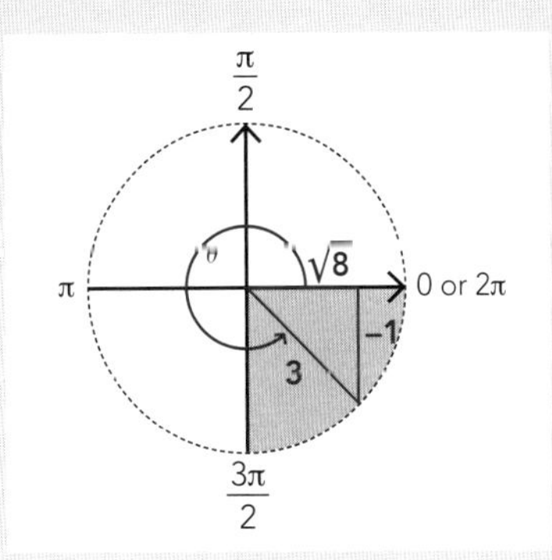

$\therefore \sin\theta = \frac{1}{3}$ or $\sin\theta = \frac{-1}{3}$

$\cos\theta = \frac{-\sqrt{8}}{3}$ $\cos\theta = \frac{\sqrt{8}}{3}$

Answer the following questions.

1 **a** If $\cos\theta = \frac{1}{2}$ and $0 < \theta < 2\pi$, find all possible values for θ.

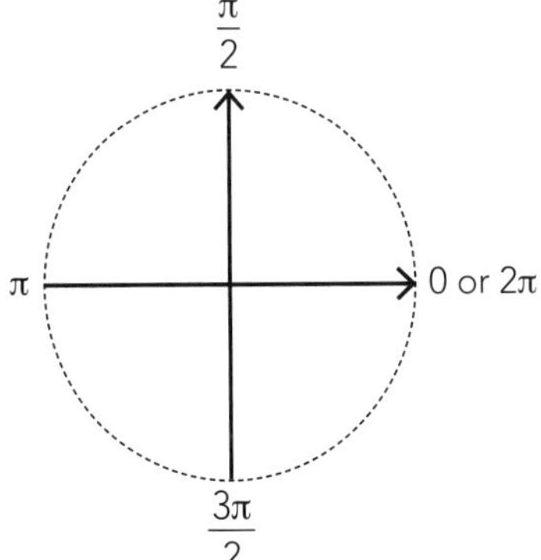

b Find exact values of $\sin\theta$ and $\tan\theta$.

2 **a** If $\tan\theta = 1$ and $0 < \theta < 2\pi$, find all possible values for θ.

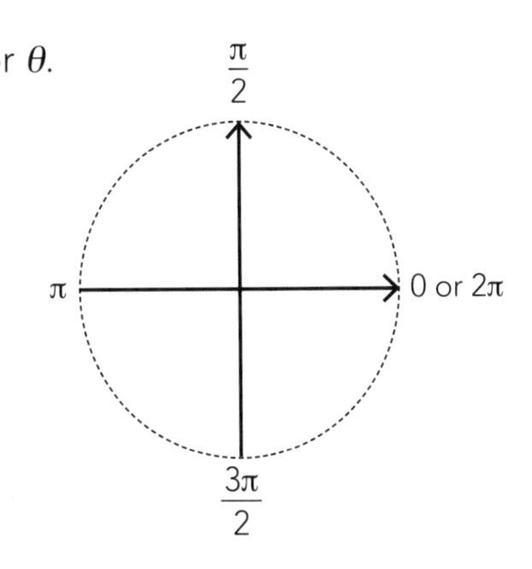

b Find exact values of $\sin\theta$ and $\cos\theta$.

3 **a** If $\cos\theta = -\frac{1}{2}$ and $0 < \theta < 2\pi$, find all possible values for θ.

b Find exact values of $\sin\theta$ and $\tan\theta$.

4 **a** If $\tan\theta = -\frac{1}{\sqrt{3}}$ and $0 < \theta < 2\pi$, find all possible values for θ.

b Find exact values of $\sin\theta$ and $\cos\theta$.

5 If $\cos\theta = \frac{4}{5}$ and $0 < \theta < 2\pi$, find exact values for $\sin\theta$ and $\tan\theta$.

ISBN: 9780170425728

6 If $\sin \theta = -\frac{1}{3}$ and $0 < \theta < 2\pi$, find exact values for $\cos \theta$ and $\tan \theta$.

7 If $\tan \theta = 1.5$ and $0 < \theta < 2\pi$, find exact values for $\cos \theta$ and $\tan \theta$.

8 If $\sin \theta = \frac{5}{12}$ and $0 < \theta < 2\pi$, find exact values for $\cos \theta$ and $\tan \theta$.

9 If $\cos \theta = -0.6$ and $0 < \theta < 2\pi$, find exact values for $\sin \theta$ and $\tan \theta$.

ISBN: 9780170425728

Transformations of trigonometric functions

- You need to be able to transform sine and cosine graphs.
- Transformations include:
  - translations – vertical, horizontal and combinations of these
  - enlargements – vertical, horizontal and combinations of these
  - combinations of translations and enlargements.

In general:

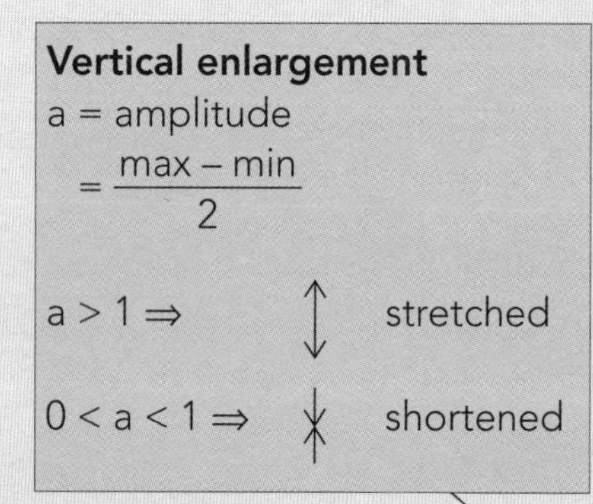

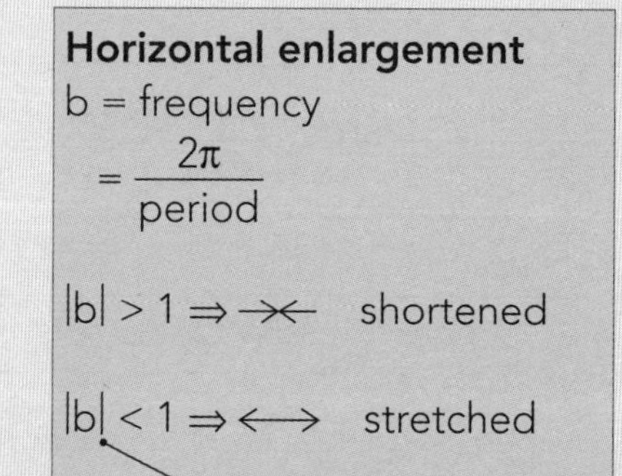

Absolute value of b = its magnitude without regard to its sign, e.g. $|-3| = 3$.

$$f(x) = a \sin b(x + c) + d$$
or
$$f(x) = a \cos b(x + c) + d$$

Horizontal translation

$+ c \Rightarrow$ shift left ⟵

$- c \Rightarrow$ shift right ⟶

Vertical translation

$d = \frac{\text{max} + \text{min}}{2}$

$+ d \Rightarrow$ shift up ↑

$- d \Rightarrow$ shift down ↓

ISBN: 9780170425728

Translations

1 Vertical translations: f(x) = sin x ± d or f(x) = cos x ± d

- A vertical translation shifts the graph **up** or **down**.
- It does **not change** the **shape**.
- You can shift a graph
  up by **adding** a number to the function
  or **down** by **subtracting** a number from a function.

$$d = \frac{\textbf{maximum value of function + minimum value of function}}{2}$$

This is the **average** of the two values.

Examples:

1 **f(x) = sin x + 0.5** — d = + 0.5 ⇒ up 0.5 or $d = \frac{1.5 + (-0.5)}{2} = +0.5$

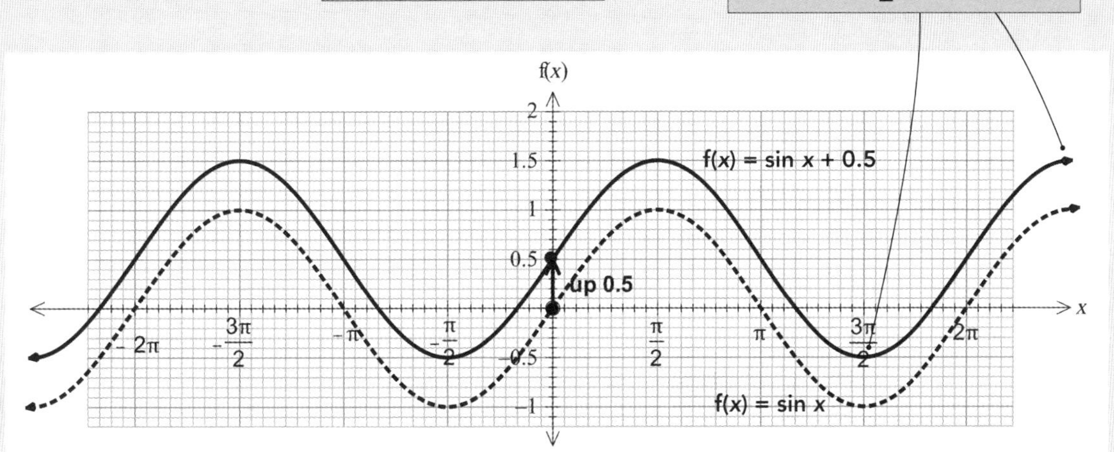

2 **f(x) = cos x – 2** — d = –2 ⇒ down 2 or $d = \frac{(-1) + (-3)}{2} = -2$

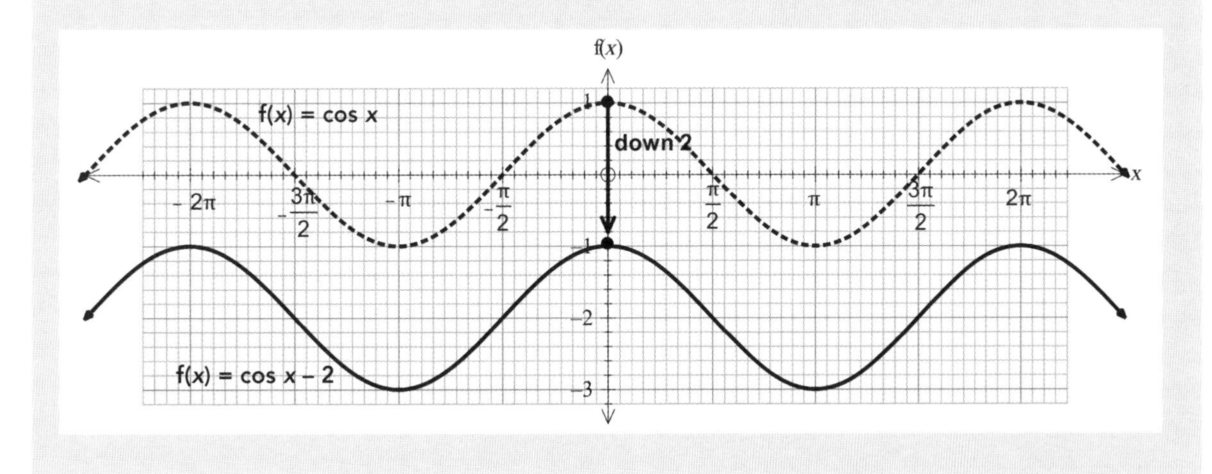

Draw the following graphs.

1 $f(x) = \cos x + 0.5$

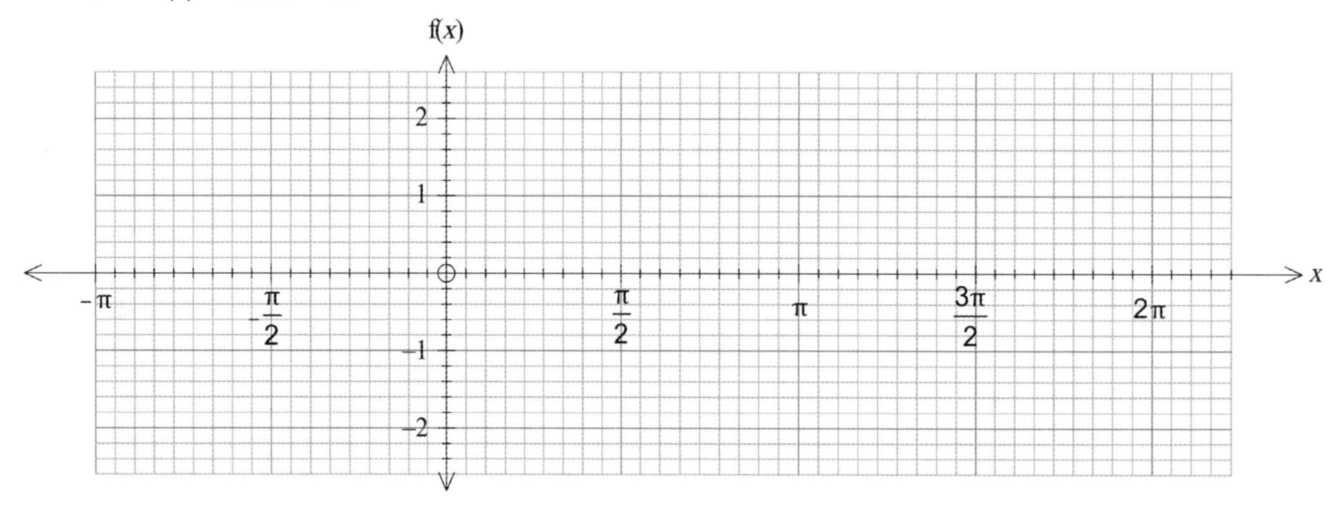

2 $f(x) = \sin x - 0.6$

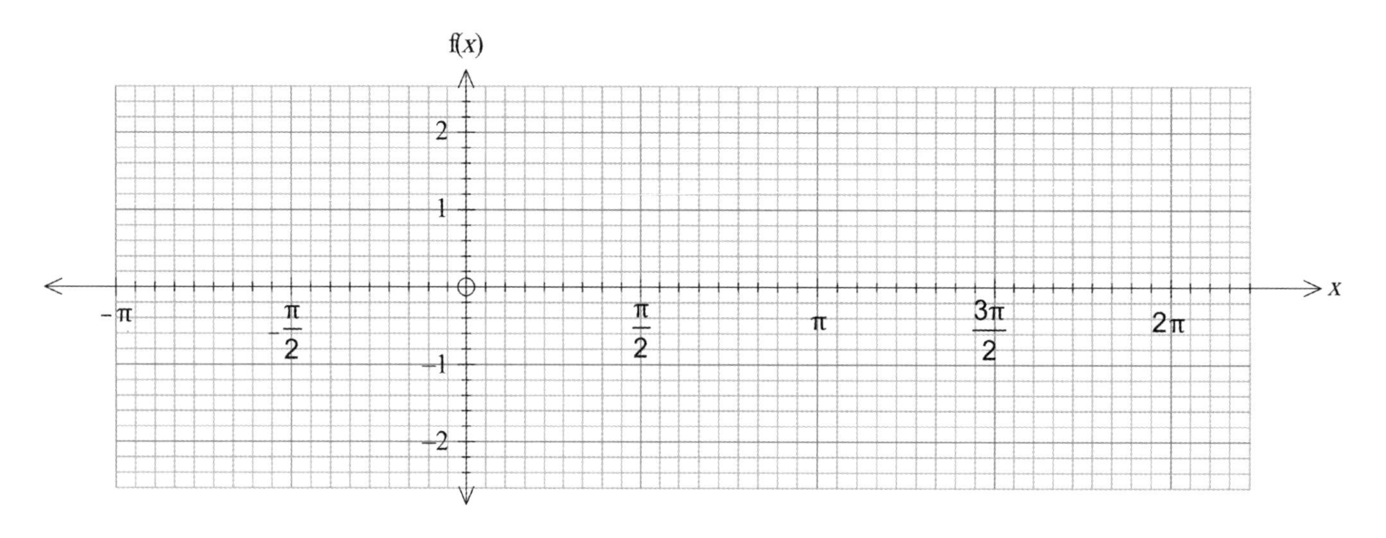

3 $f(x) = \cos x - 2.6$

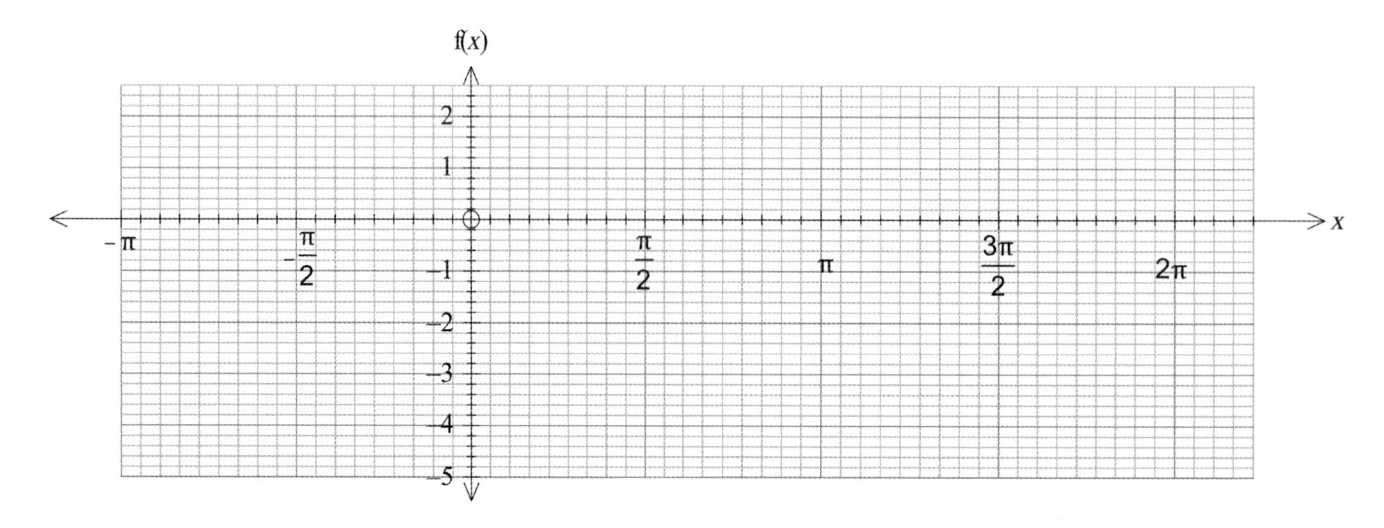

ISBN: 9780170425728

2 Horizontal translations: $f(x) = \sin(x \pm c)$ or $f(x) = \cos(x \pm c)$

- A horizontal translation shifts the graph **left** or **right**.
- It does **not change** the **shape**.
- You can shift a graph
  left by **adding** a number to the x term
  or **right** by **subtracting** a number from the x term.

Notice that the graphs move in the **opposite** direction from what you might expect.

Examples:

Be careful – this is **not** the same as $f(x) = \cos x - \frac{\pi}{2}$

1 $\mathbf{f(x) = \cos\left(x - \frac{\pi}{2}\right)}$

$-\frac{\pi}{2} \Rightarrow$ **right** $\frac{\pi}{2}$

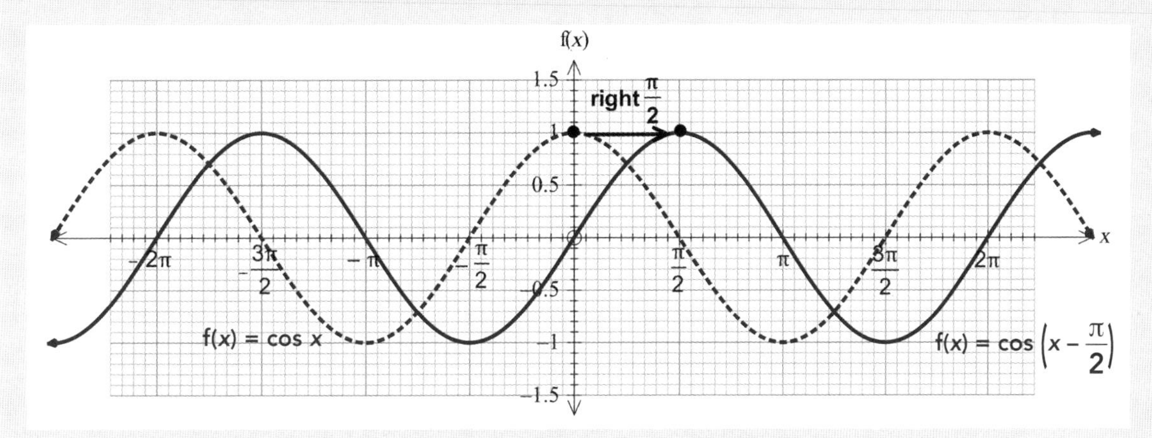

Notice that $f(x) = \cos\left(x + \frac{3\pi}{2}\right)$ would produce an identical graph.

2 $\mathbf{f(x) = \sin\left(x + \frac{4\pi}{3}\right)}$

$+\frac{4\pi}{3} \Rightarrow$ **left** $\frac{4\pi}{3}$

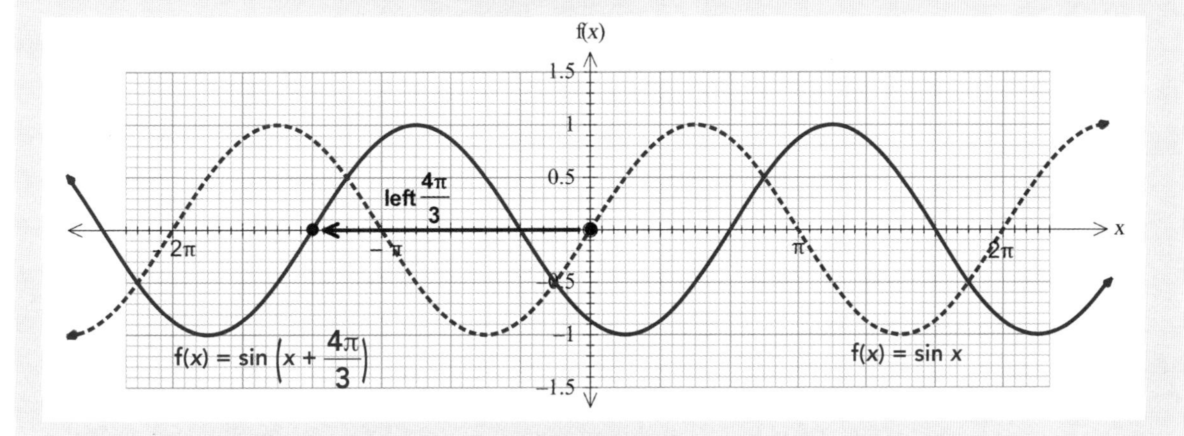

Notice that $f(x) = \sin\left(x - \frac{2\pi}{3}\right)$ would produce an identical graph.

Draw the following graphs.

1 $f(x) = \sin\left(x - \frac{5\pi}{4}\right)$

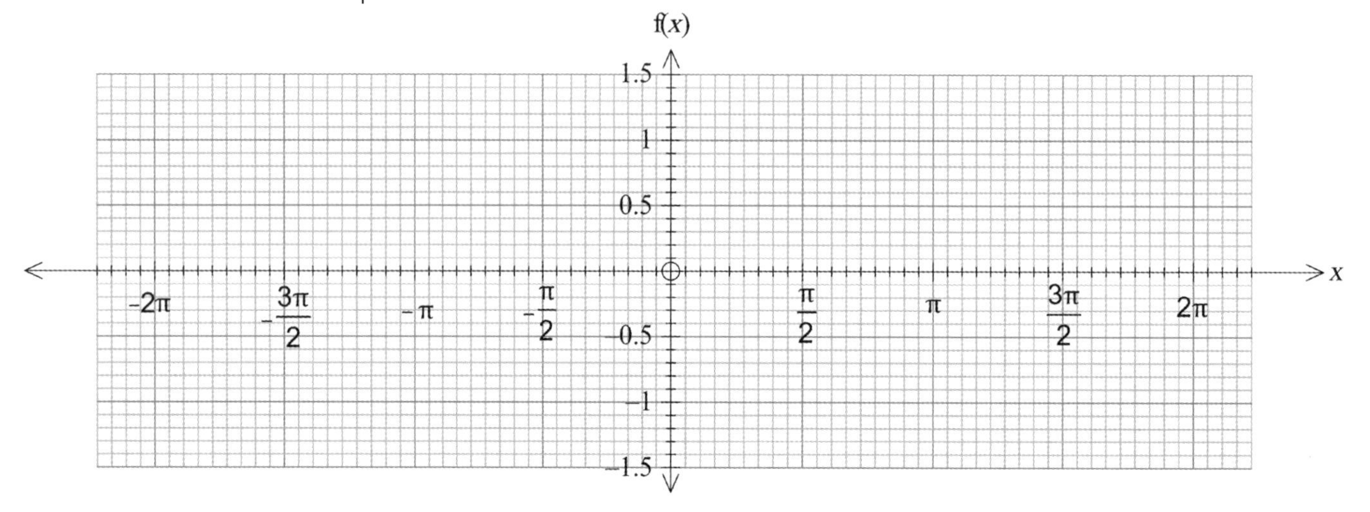

2 $f(x) = \cos\left(x + \frac{\pi}{3}\right)$

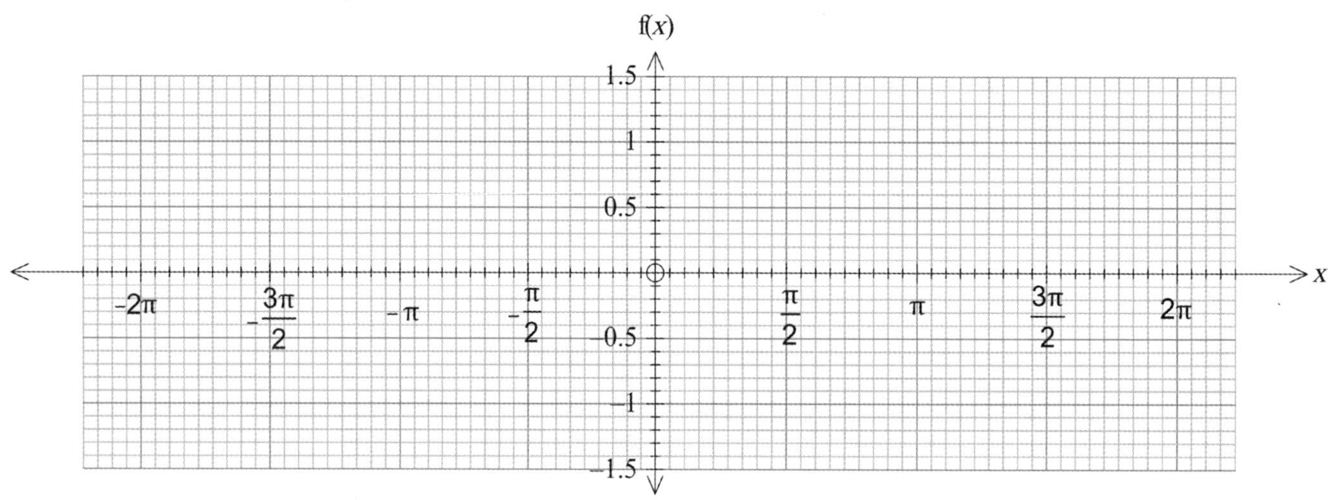

3 $f(x) = \cos\left(x - \frac{5\pi}{6}\right)$

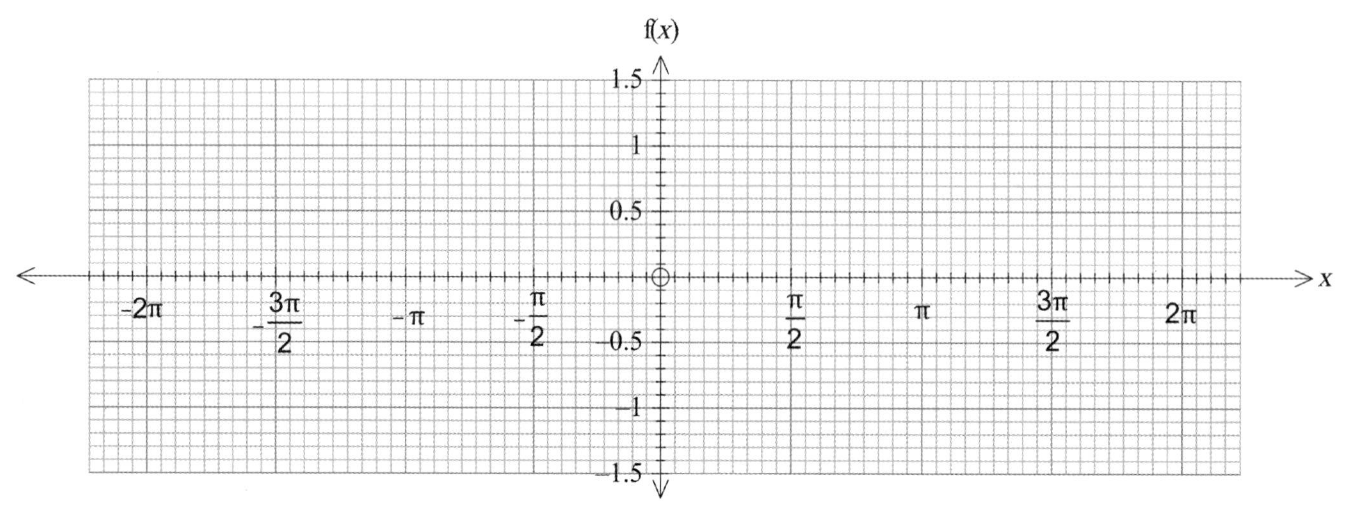

ISBN: 9780170425728

3 Combinations of translations

a Drawing combinations of translations

Examples:

1 $f(x) = \cos(x + \pi) + 1$

$+ 1 \Rightarrow$ up 1

$+ \pi \Rightarrow$ left π

Draw the function $f(x) = \cos x$ and translate every point left by π and up 1.

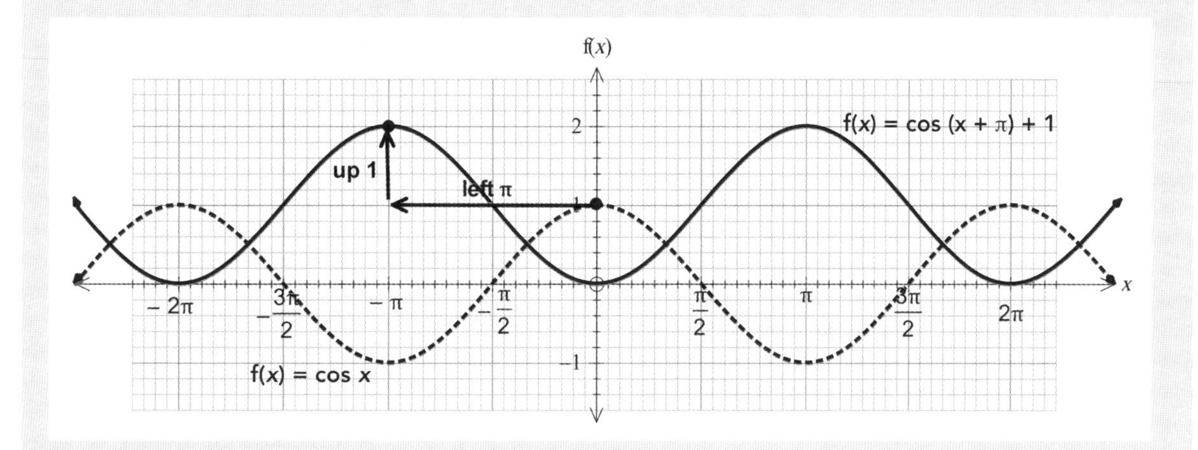

2 $f(x) = \sin\left(x - \frac{\pi}{4}\right) - 1.5$

$- 1.5 \Rightarrow$ down 1.5

$-\frac{\pi}{4} \Rightarrow$ right $\frac{\pi}{4}$

Draw the function $f(x) = \sin x$ and translate every point right by $\frac{\pi}{4}$ and down 1.5.

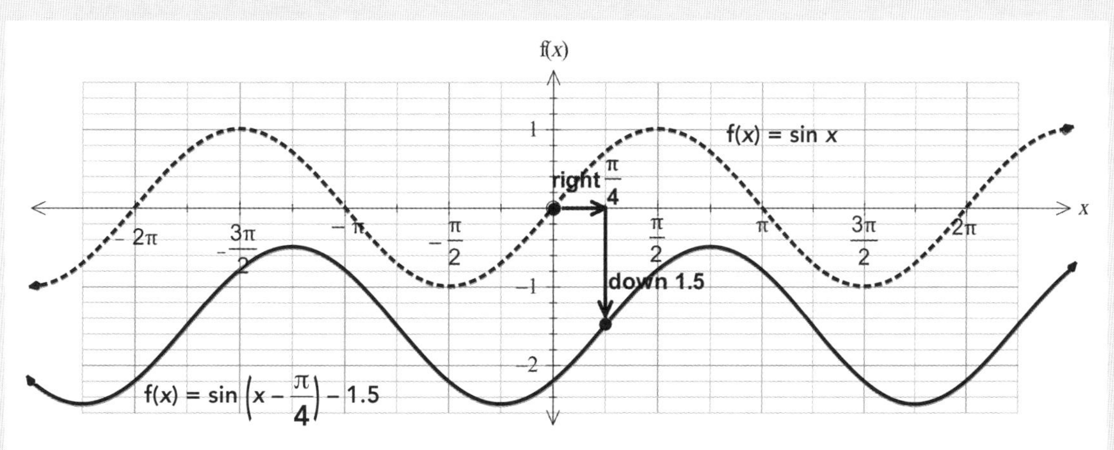

ISBN: 9780170425728

b Writing equations for combinations of translations

Example 1: For a sine curve
Step 1: Calculate the value of d

$$d = \frac{\text{maximum value of function + minimum value of function}}{2}$$

Draw the horizontal axis of the function.

The horizontal axis of the function

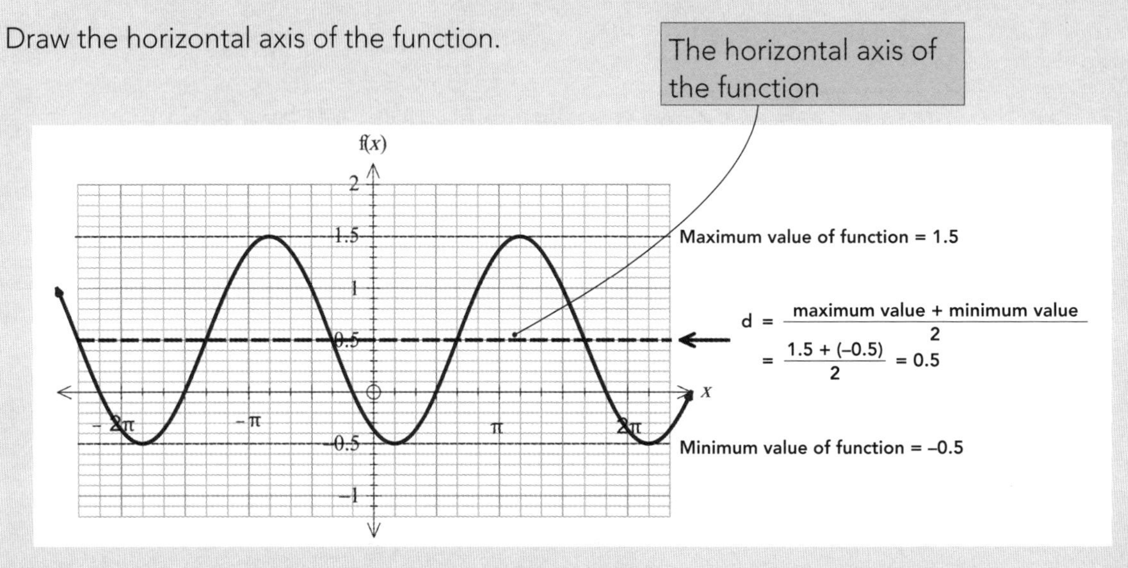

$\therefore$ **d = +0.5**

Step 2: Calculate the value of c

c = horizontal distance between the f(x) axis and an *increasing midpoint* on the function

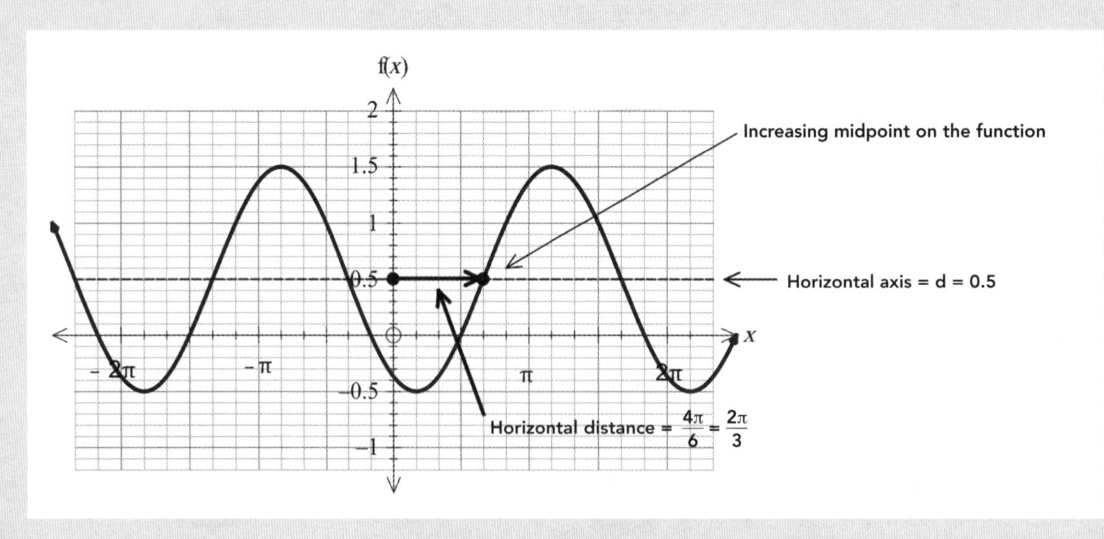

$$\therefore c = -\frac{2\pi}{3}, \text{ so } f(x) = \sin\left(x - \frac{2\pi}{3}\right) + 0.5$$

Don't forget that horizontal translations take the **opposite sign** of the direction.

 ISBN: 9780170425728

An alternative value for c

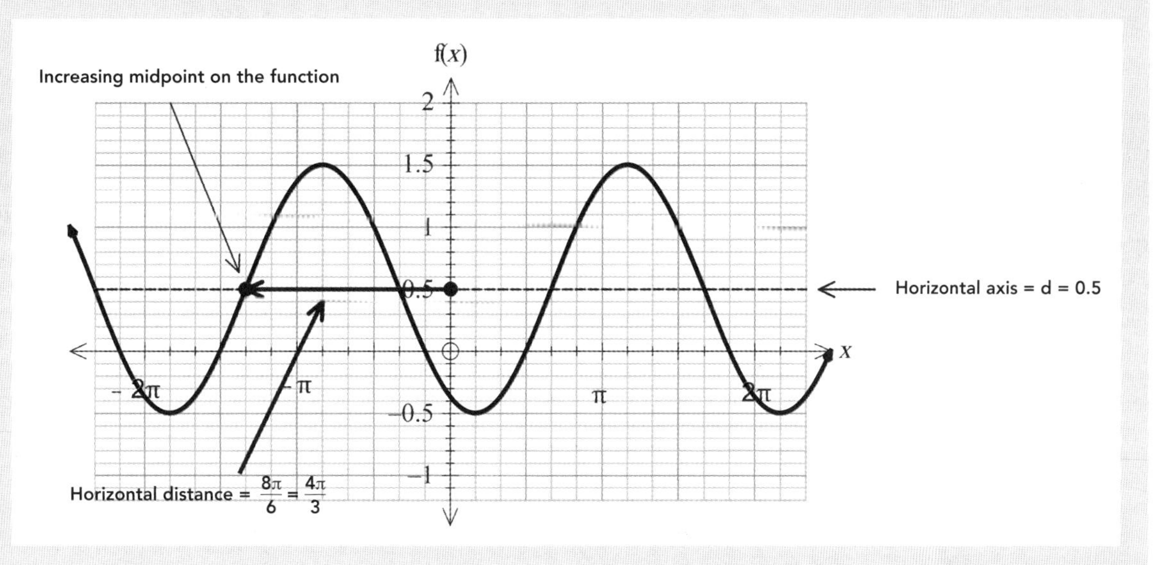

$$\therefore c = +\frac{4\pi}{3}, \text{ so } f(x) = \sin\left(x + \frac{4\pi}{3}\right) + 0.5$$

Example 2: For a cosine curve

Step 1: Calculate the value of d

$$d = \frac{\text{maximum value of function + minimum value of function}}{2}$$

Draw the horizontal axis of the function.

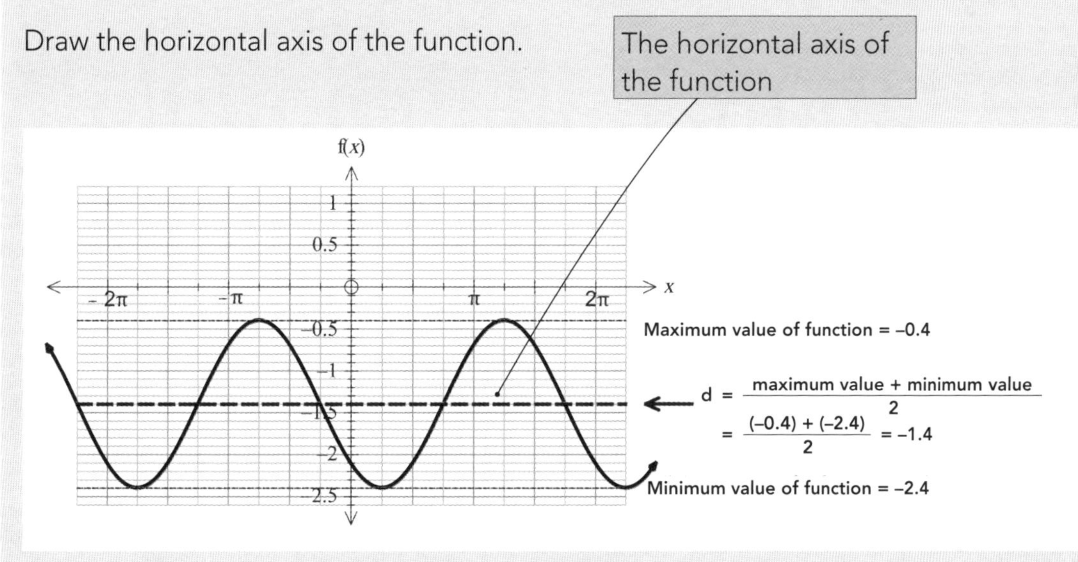

$$\therefore d = -1.4$$

Step 2: Calculate the value of c

c = horizontal distance between the f(x) axis and a *maximum point* on the function

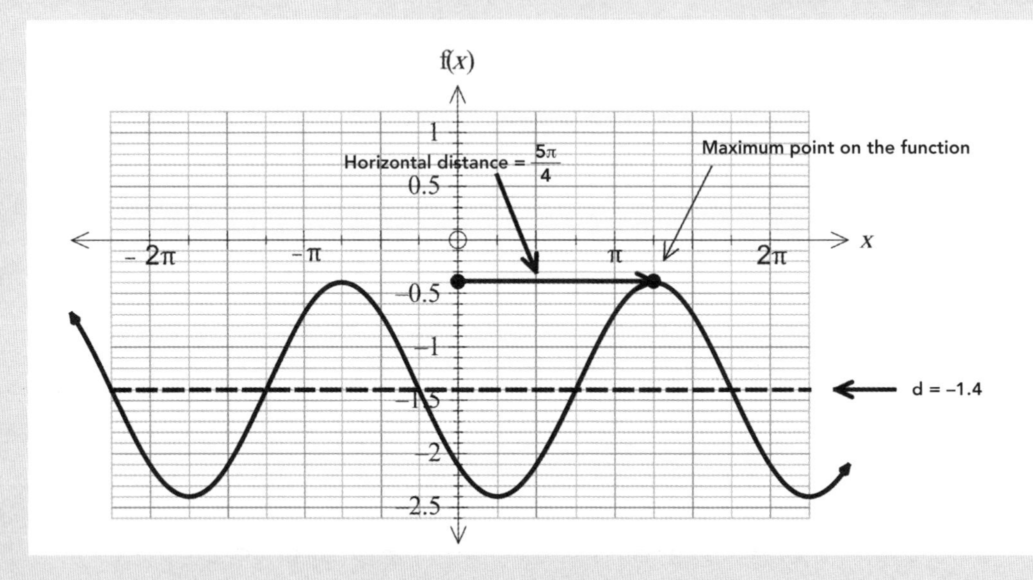

$$\therefore c = -\frac{5\pi}{4}, \text{ so } f(x) = \cos\left(x - \frac{5\pi}{4}\right) - 1.4$$

An alternative value for c

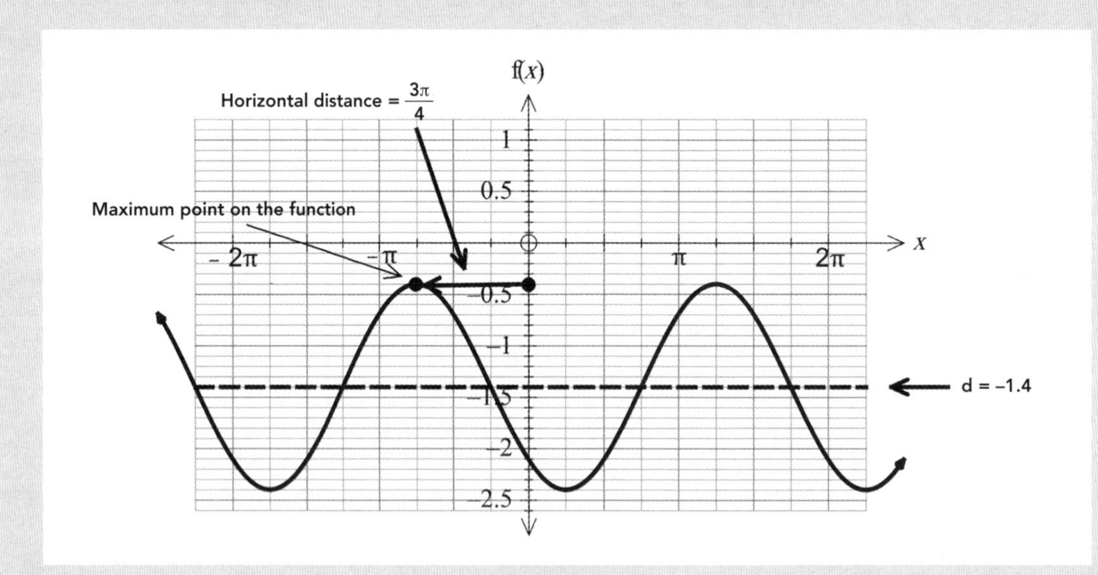

$$\therefore c = \frac{3\pi}{4}, \text{ so } f(x) = \cos\left(x + \frac{3\pi}{4}\right) - 1.4$$

ISBN: 9780170425728

Match the following graphs to the equations below.
Hint for 1–4: Consider where the point (0, 1) on $f(x) = \cos x$ has moved to.

1

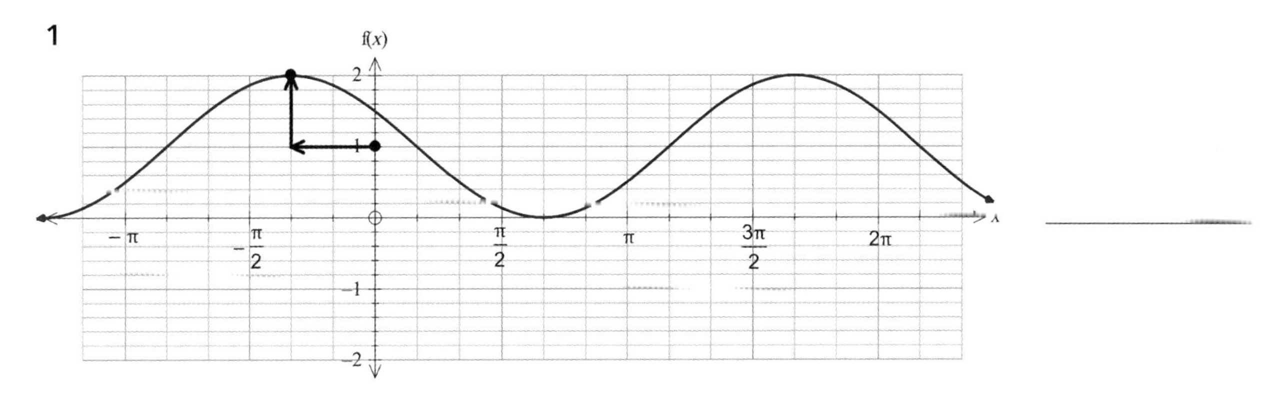

____

2

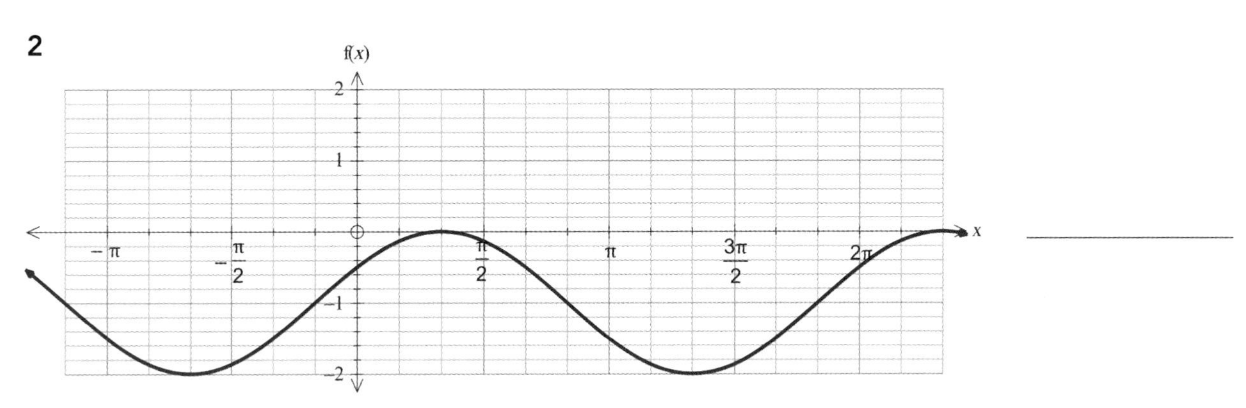

____

3

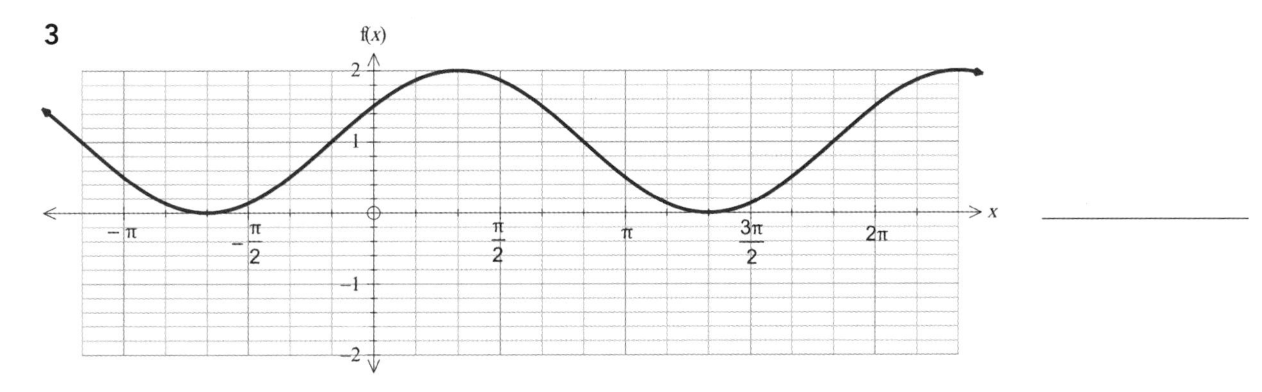

____

4

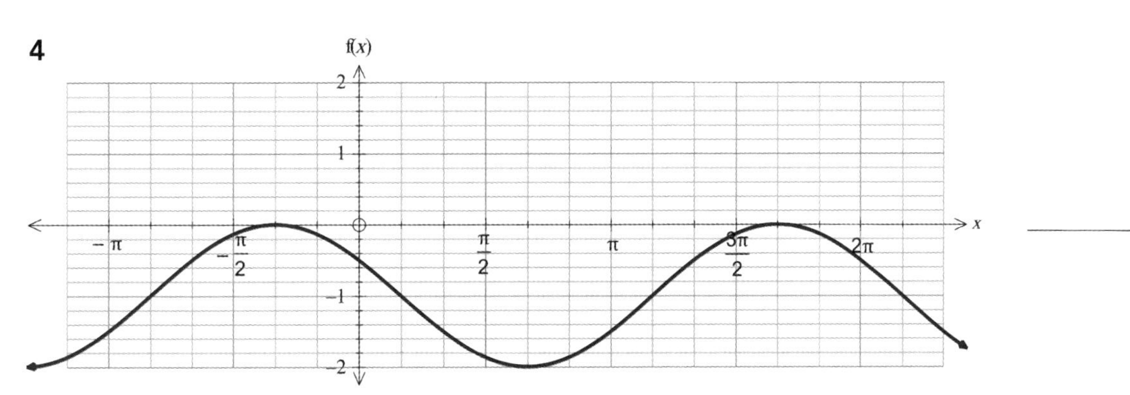

____

A $f(x) = \cos\left(x - \frac{\pi}{3}\right) - 1$

B $f(x) = \cos\left(x - \frac{5\pi}{3}\right) - 1$

C $f(x) = \cos\left(x + \frac{\pi}{3}\right) + 1$

D $f(x) = \cos\left(x - \frac{\pi}{3}\right) + 1$

ISBN: 9780170425728

Write equations for the following functions.

5 A vertical translation and vertical enlargement of $f(x) = \sin x$ which has a maximum value of 4 and a minimum value of –1.

6 A vertical translation and vertical enlargement of $f(x) = \cos x$ which has a maximum value of 2.4 and a minimum value of –8.2.

Draw graphs of the following functions.

7 $f(x) = \cos\left(x - \frac{3\pi}{4}\right) + 0.6$

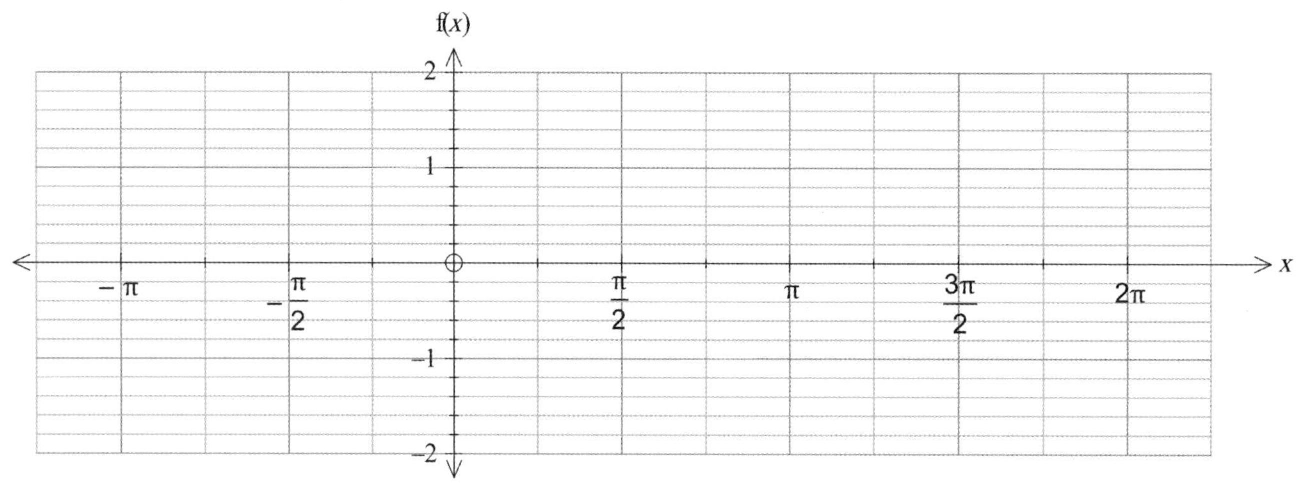

8 $f(x) = \sin\left(x + \frac{7\pi}{6}\right) - 0.5$

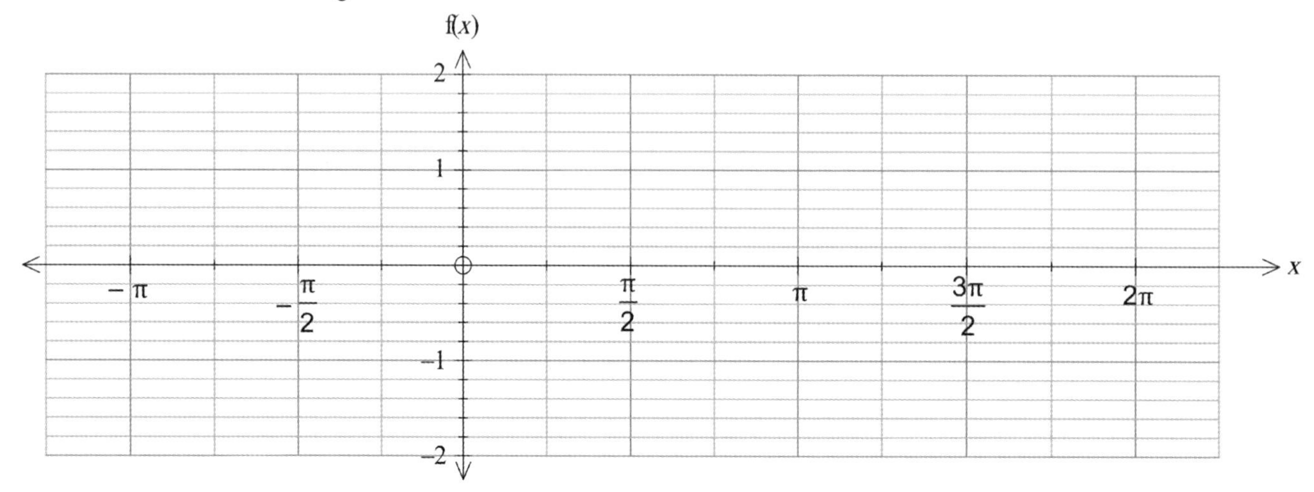

 ISBN: 9780170425728

Write equations for the following graphs.

9 This graph is a function of f(x) = sin *x*.
Hint: Consider where the point (0, 0) on f(x) = sin *x* has moved to.

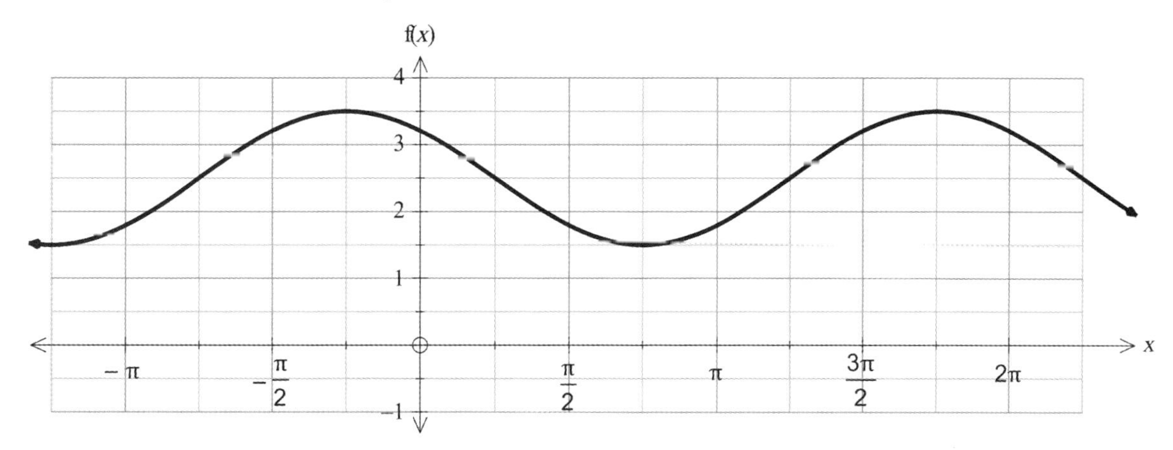

10 This graph is a function of f(x) = cos x.
Hint: Consider where the point (0, 1) on f(x) = cos *x* has moved to.

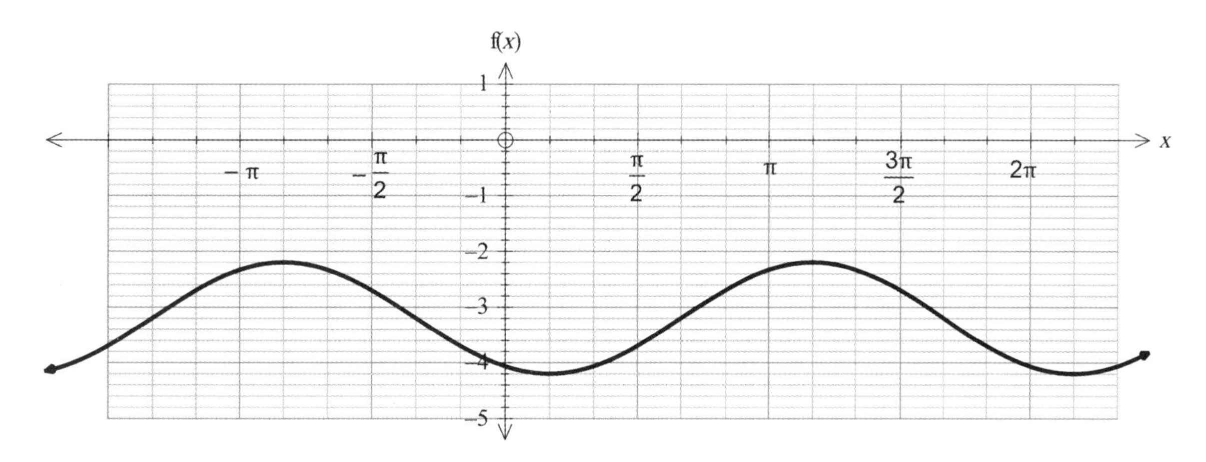

11 This graph is a function of f(x) = cos x.

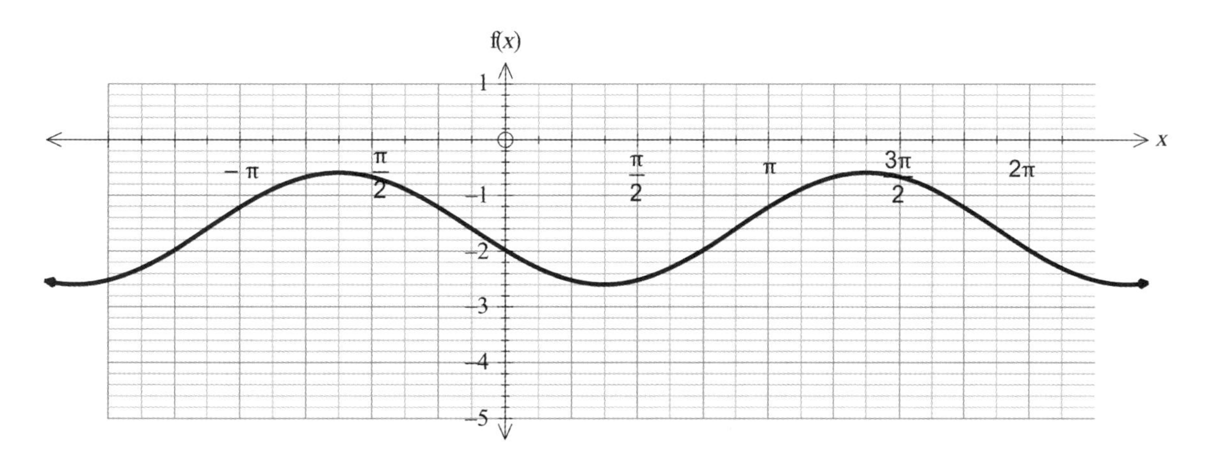

Enlargements

- Enlargements can occur in a **vertical** or **horizontal** direction.
- Remember, 'enlargement' is a general term, which may result in the curve being **stretched** or **shrunk** in one direction.

1 Vertical enlargements: f(x) = asin *x* or f(x) = acos *x*

- These cause stretching or shortening along the **y** or **f(x)** axis.
- They change the **amplitude**.
- You can enlarge a function along the *y* or f(x) axis by **multiplying the entire function by a number (a)**.

a **>** 1 ⇒ function is **stretched** vertically ↕ 0 < a **<** 1 ⇒ function is **shortened** vertically

Examples:

1 f(x) = 0.5sin x

0.5 ⇒ shortening along the **f(x) axis**. The f(x) values are **halved**.

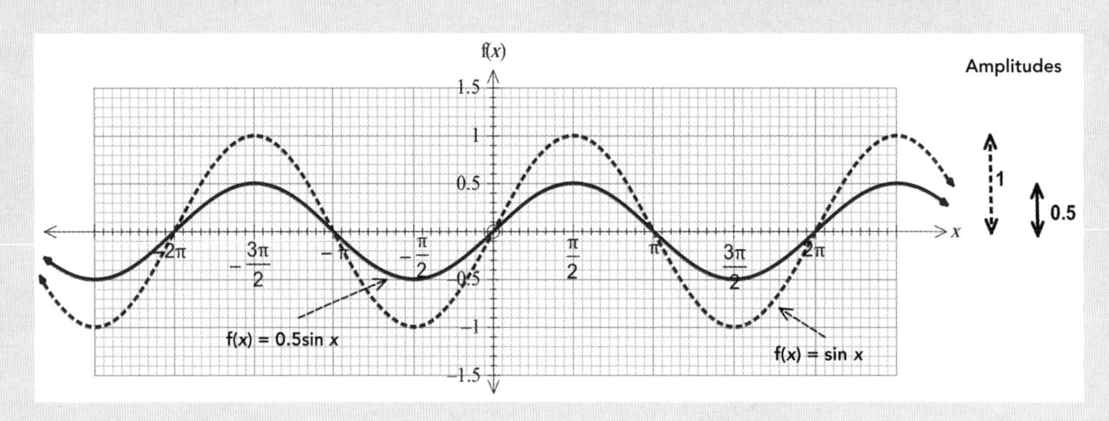

Period = 360°
Amplitude = 0.5

2 f(x) = 1.5cos x

1.5 ⇒ stretching along the **f(x) axis**. The f(x) values are **multiplied by 1.5**.

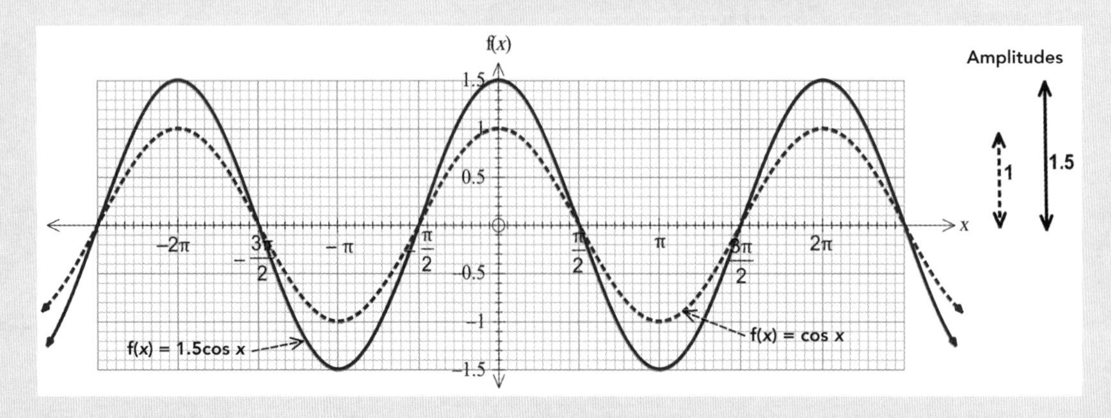

Period = 360°
Amplitude = 1.5

ISBN: 9780170425728

Draw the following graphs.

1 f(x) = 2sin x

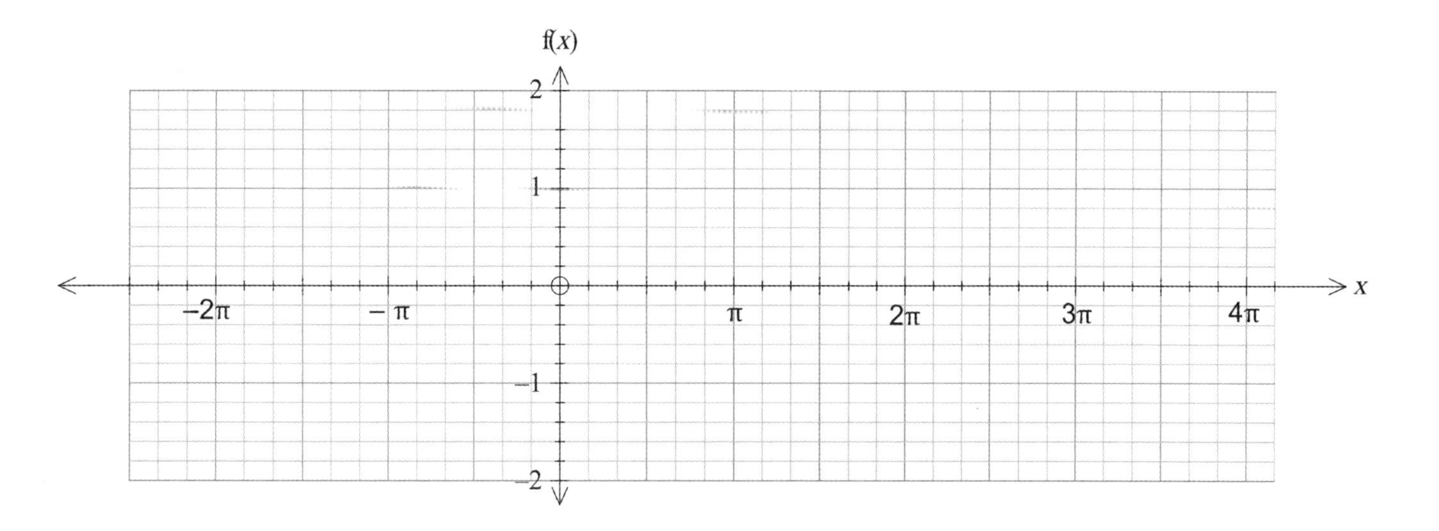

2 f(x) = 1.6cos x

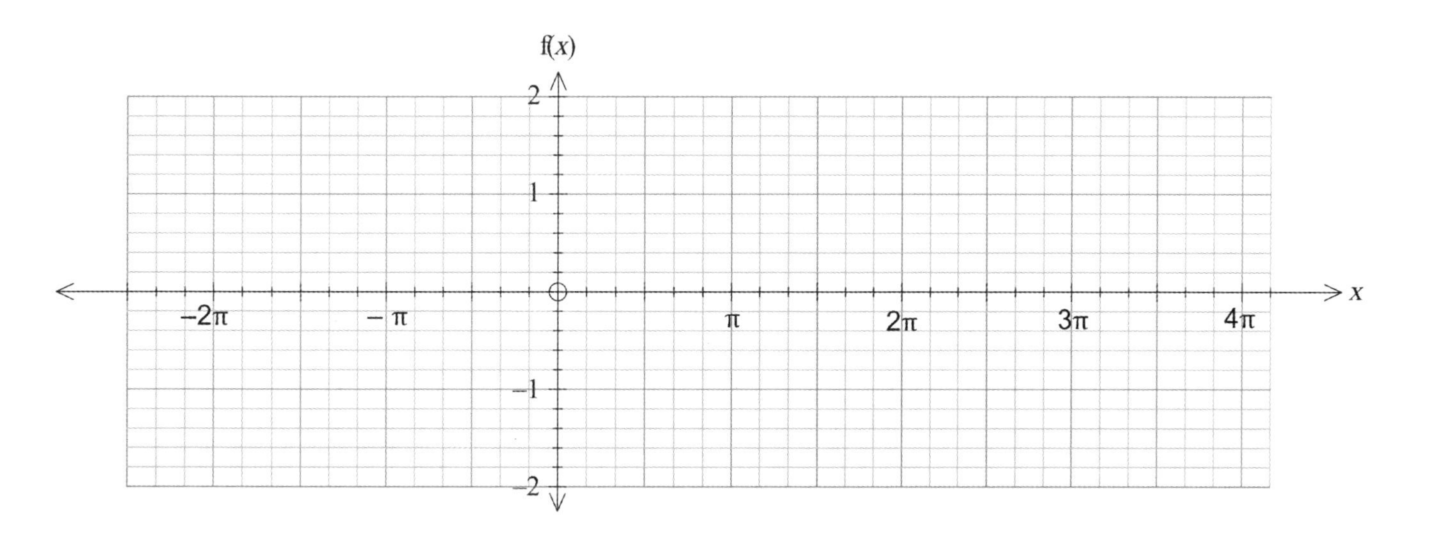

ISBN: 9780170425728

2 Horizontal enlargements: $f(x) = \sin bx$ or $f(x) = \cos bx$

- These cause **stretching** or **shortening** along the **x**-axis.
- They change the **period**.
- You can enlarge a function along the x-axis by **multiplying x by a number (b)**.

$b > 1 \Rightarrow$ function is **shortened** horizontally →←

$b < 1 \Rightarrow$ function is **stretched** horizontally ←→

Notice that these are the opposite of the changes for vertical enlargements.

Examples:

1 $f(x) = \sin 0.5x$

0.5 ⇒ stretching along the **x-axis**. The x values are **doubled**.

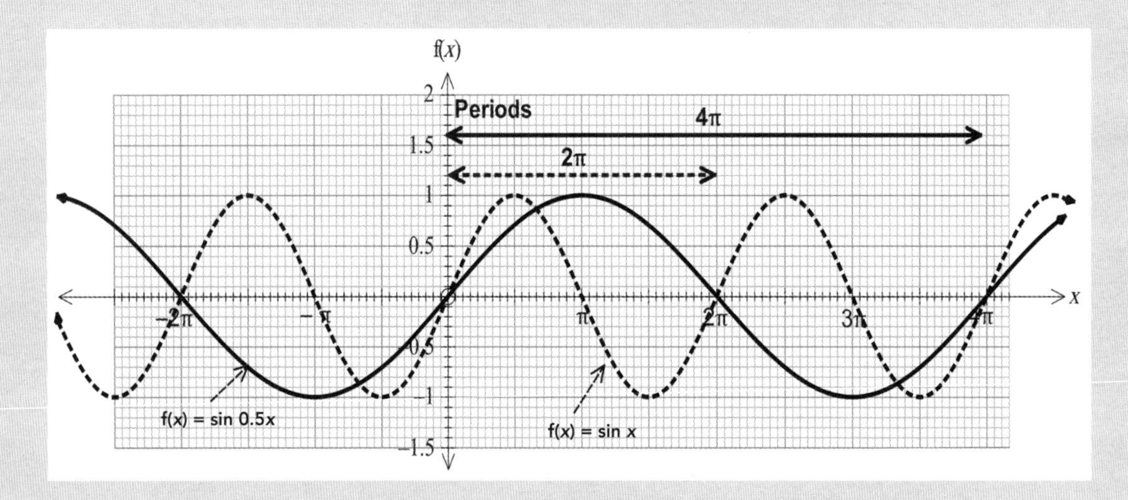

2 $f(x) = \cos 3x$

3 ⇒ shortening along the **x-axis**. The x values are **divided by 3**.

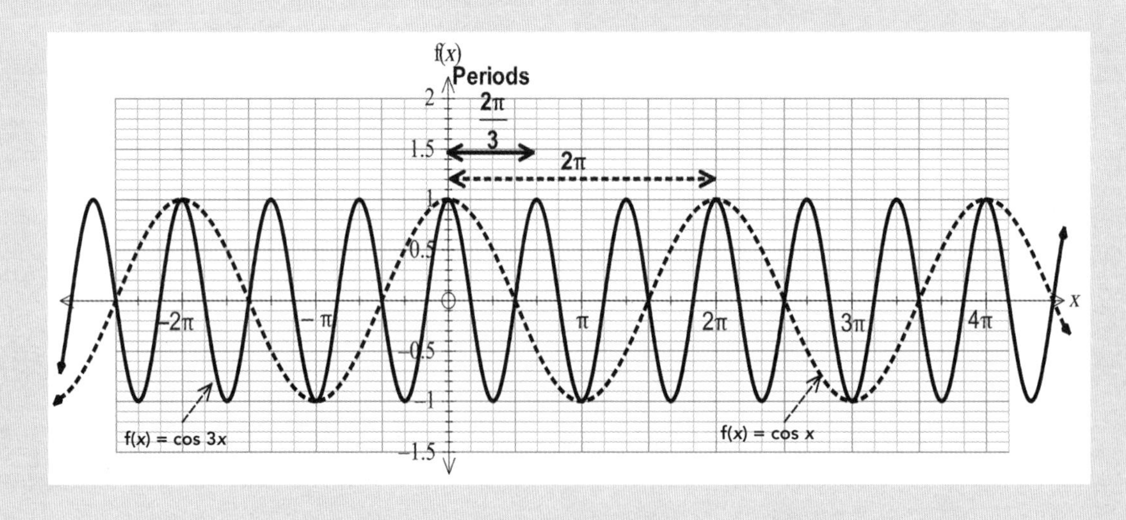

 ISBN: 9780170425728

Draw the following graphs.

1 $f(x) = \sin 2x$

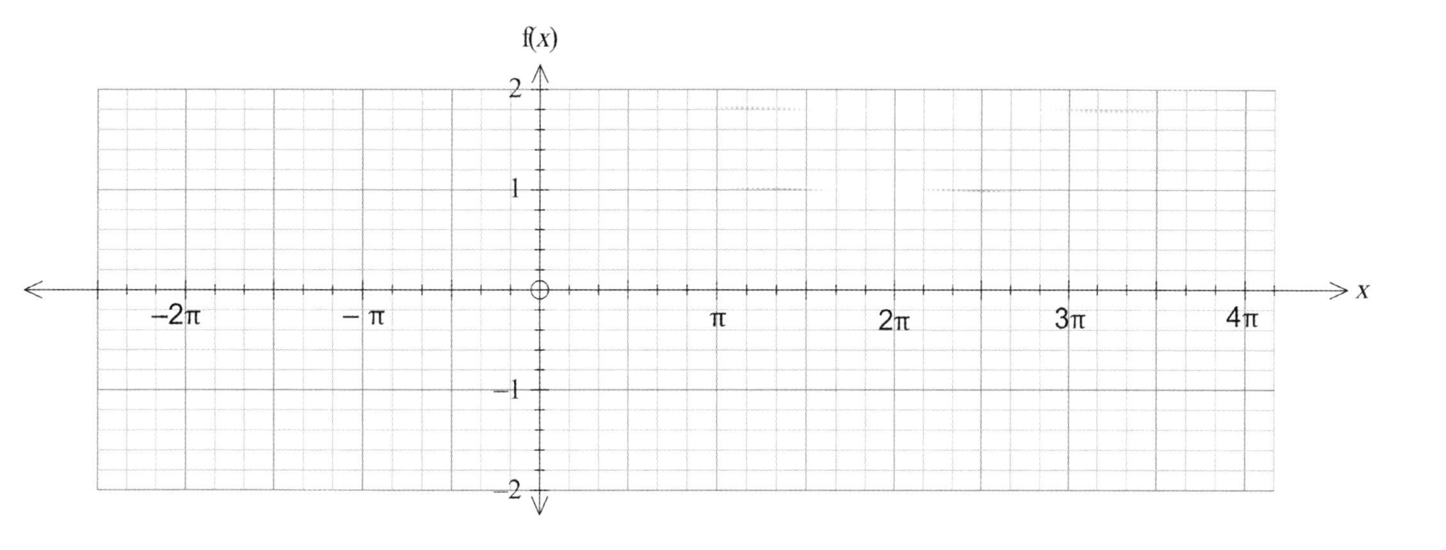

2 $f(x) = \cos \frac{1}{3}x$

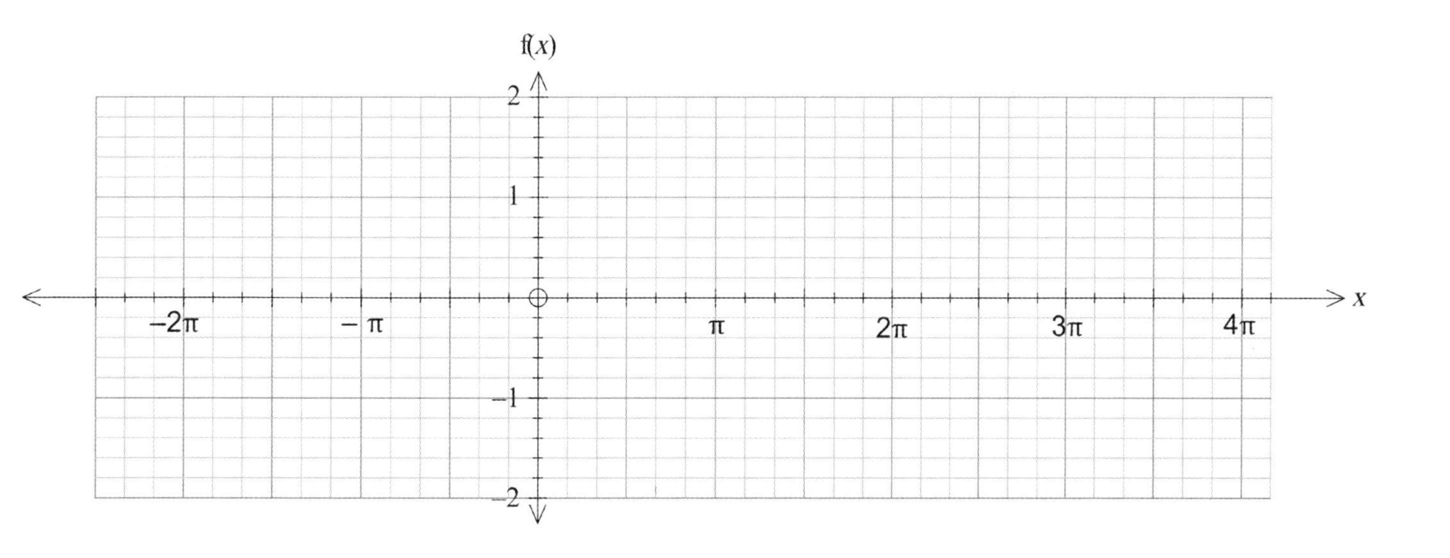

3 Combinations of enlargements

Match the following graphs to the equations below.

Hint for 1–5: Using f(x) = sin x as a basis, compare the period and the amplitude.

1

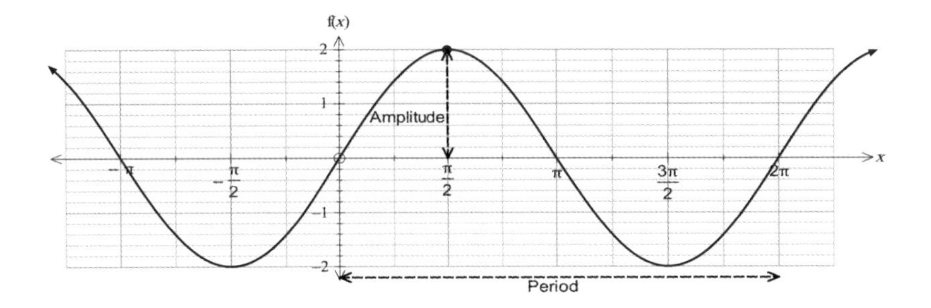

____

2

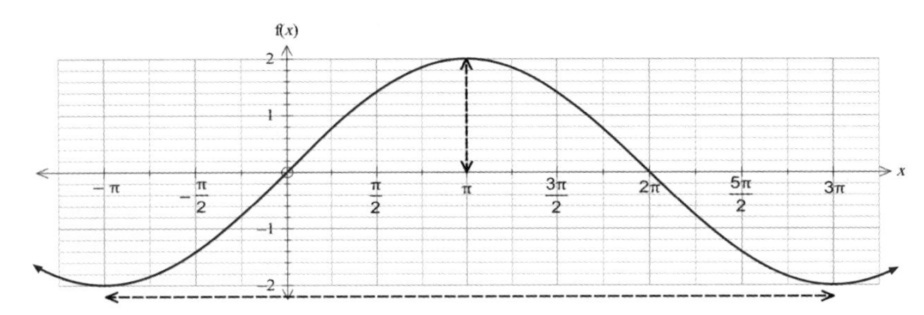

____

3

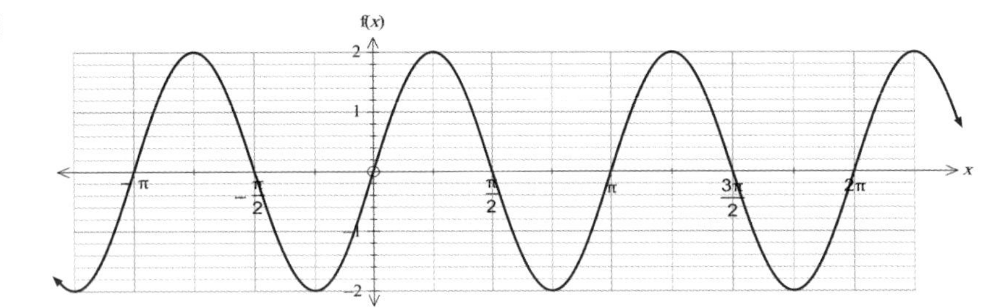

____

4

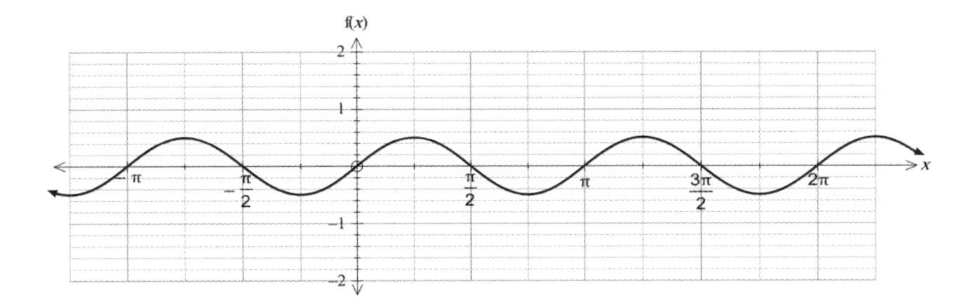

____

5

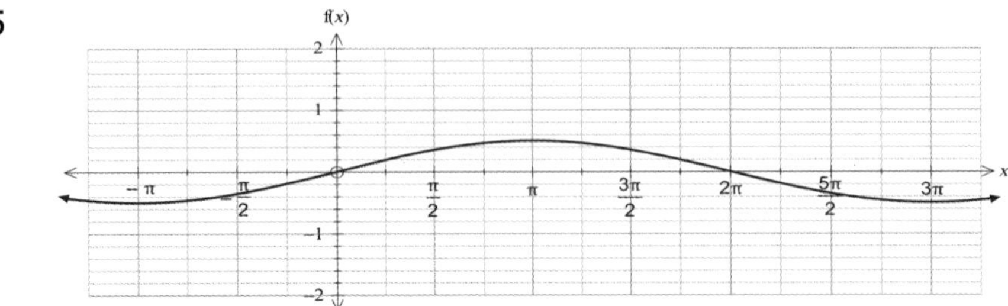

____

A f(x) = 2sin 0.5x

B f(x) = 0.5sin 2x

C f(x) = 0.5sin 0.5x

D f(x) = 2sin x

E f(x) = 2sin 2x

ISBN: 9780170425728

Write equations for the following graphs.
Hint: Find the amplitude and the period for each graph.

6

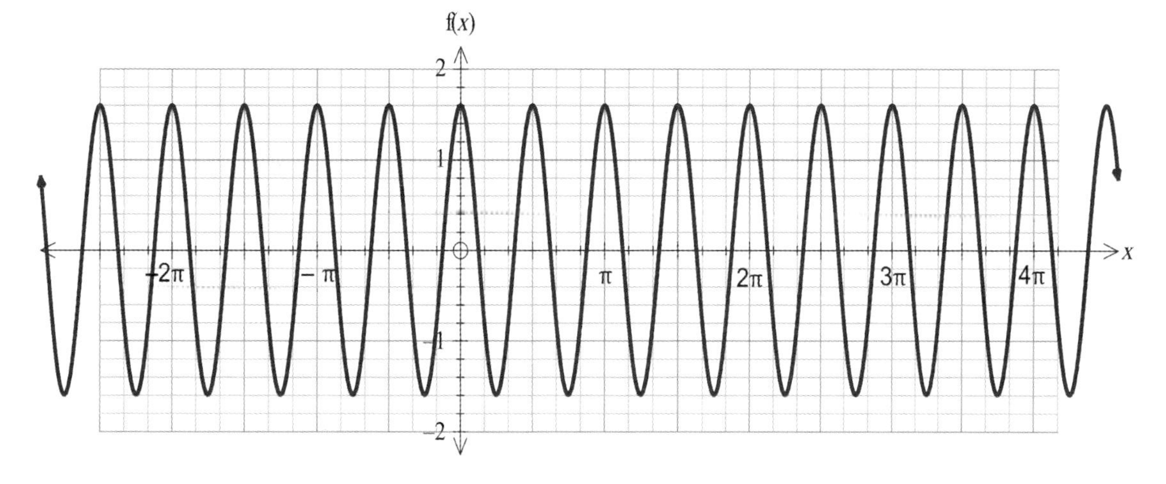

7

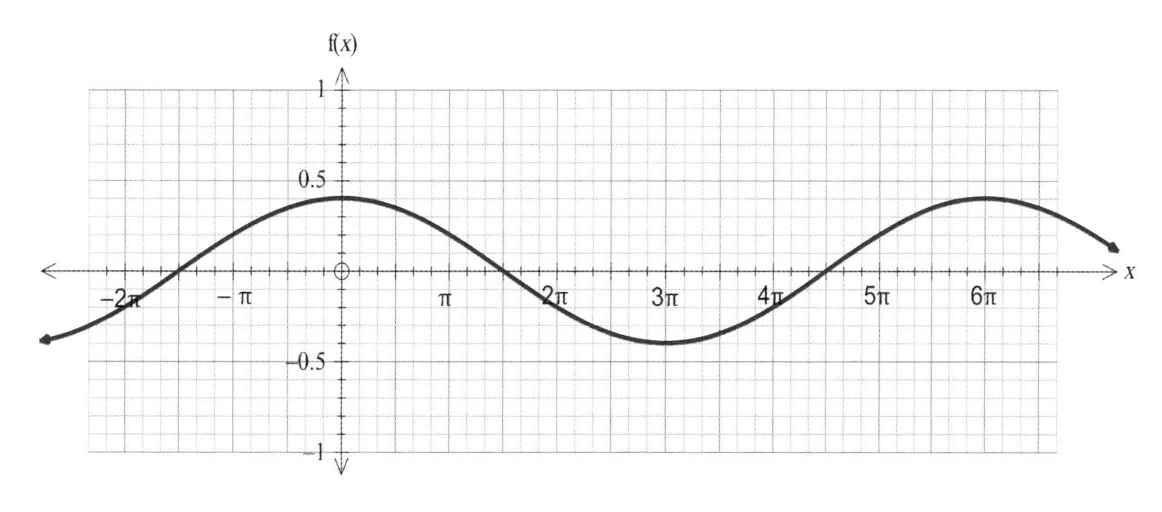

8

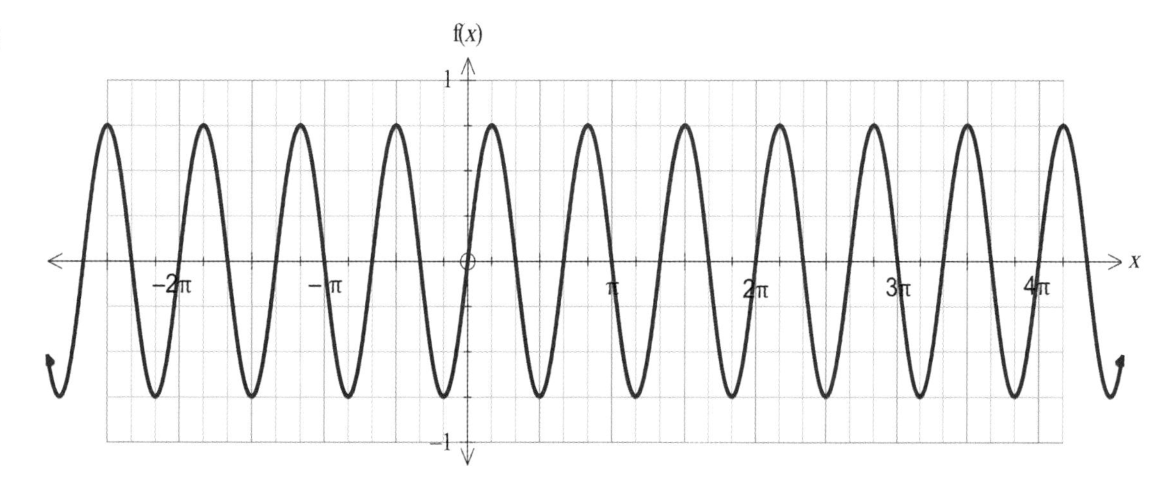

ISBN: 9780170425728

Putting it together

Finding the equation from a graph

Step 1: Find **d**, the **horizontal axis of the function**.

$$d = \frac{\text{maximum value of function} + \text{minimum value of function}}{2}$$

Draw this in.

So f(x) = acos b(x + c) – 2.6

Step 2: Find **a**, the **amplitude**.

a = maximum value – d

or

$$a = \frac{\text{maximum value of function} - \text{minimum value of function}}{2}$$

So f(x) = 1.4cos b(x + c) – 2.6

Step 3: Find **b**, the **frequency**.

$$b = \text{frequency} = \frac{2\pi}{\text{period}}$$

So f(x) = 1.4cos 3(x + c) – 2.6

Example: This is a **cosine** graph.

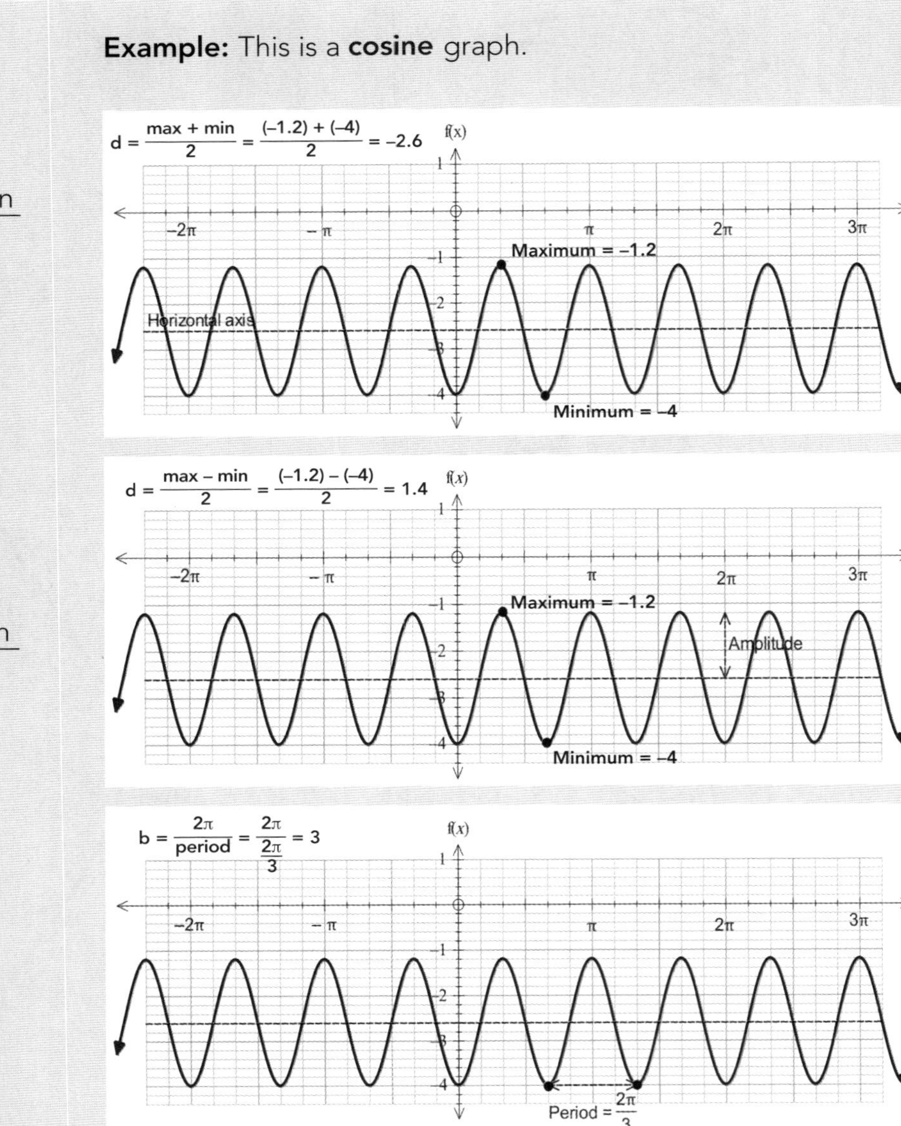

 ISBN: 9780170425728

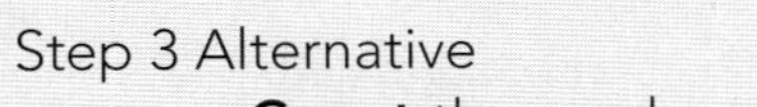

Step 3 Alternative

Count the number of cycles in 2π.

So f(x) = 1.4cos 3(x + c) – 2.6

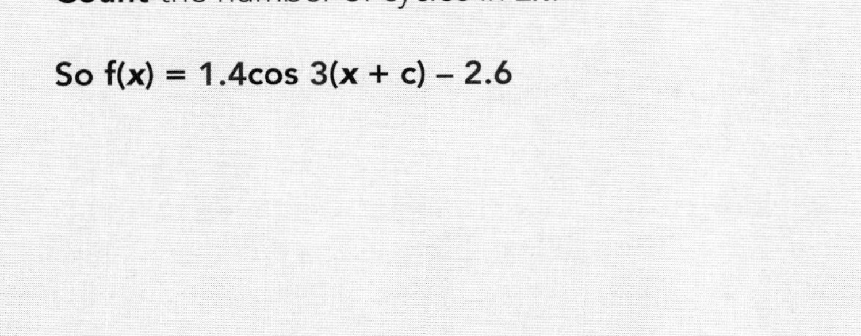

Step 4: Find **c**, the **horizontal shift**.

This is a **cosine curve**, so c = horizontal distance between the f(x) axis and the nearest **maximum** of the function.

So $f(x) = 1.4\cos 3(x - \frac{\pi}{3}) - 2.6$

Step 4: If this was a **sine curve**:

c = horizontal distance between the f(x) axis and the nearest **increasing midpoint** of the function.

So $f(x) = 1.4\sin 3(x - \frac{\pi}{6}) - 2.6$

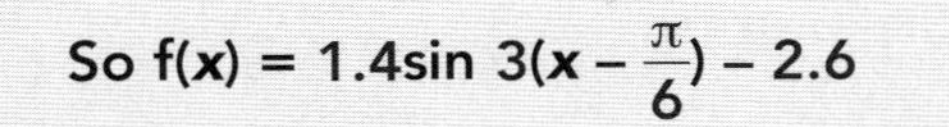

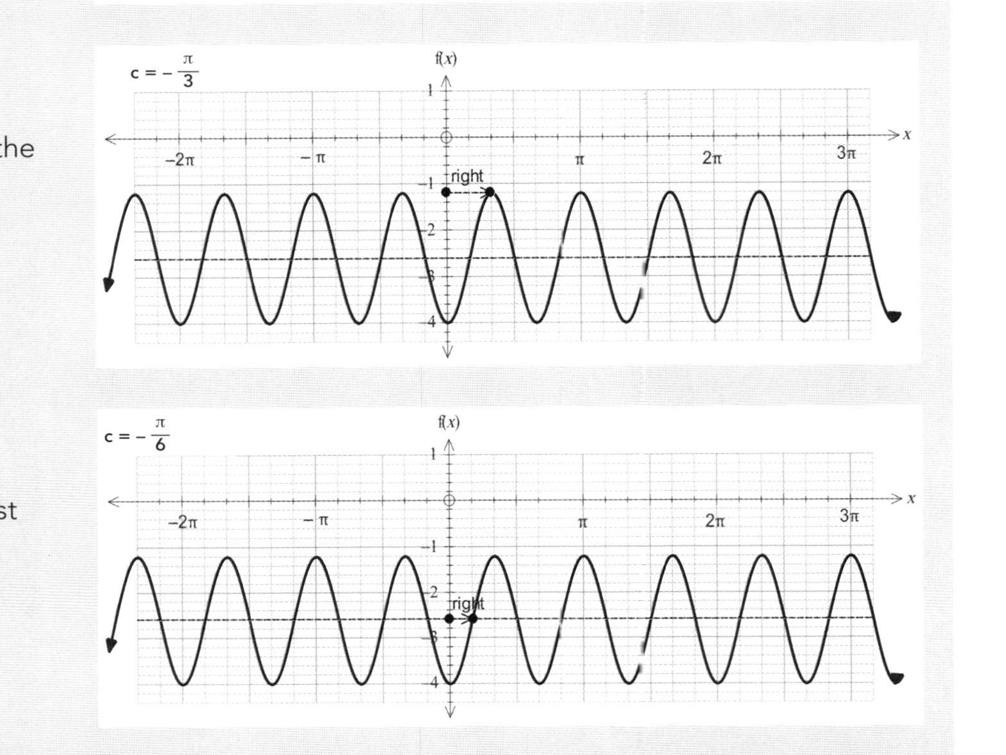

ISBN: 9780170425728

Drawing a graph from an equation

Step 1: **d** tells you where the **horizontal axis of the function** is.

$$f(x) = 1.5\sin 2(x - \frac{3\pi}{4}) + 3$$

d = +3

Draw this in.

Step 2: **a** tells you the **amplitude**.

$$f(x) = 1.5\sin 2(x - \frac{3\pi}{4}) + 3$$

a = 1.5

Draw lines for the **maximum** and **minimum** values of the function.

Step 3: **c** represents the **horizontal translation**.
Sine function: this mark must be on the horizontal axis of the function.

$$f(x) = 1.5\sin 2(x - \frac{3\pi}{4}) + 3$$

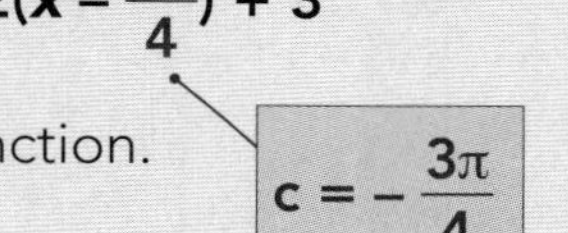

Call this the **'centre'** of the function.
Mark where this is.

Example: $f(x) = 1.5\sin 2(x - \frac{3\pi}{4}) + 3$

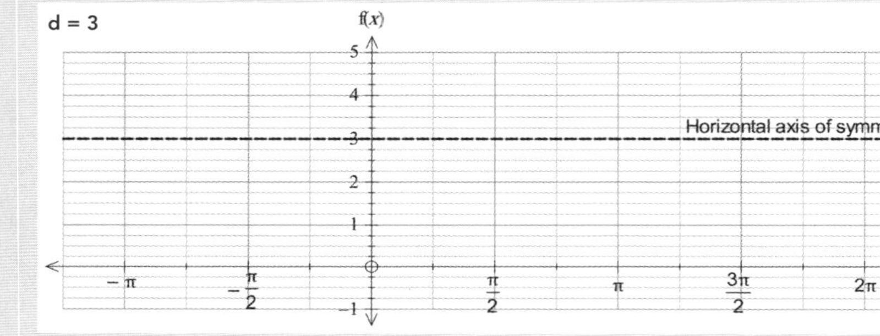

a = 1.5

f(x)

Maximum

1.5

1.5

Minimum

x

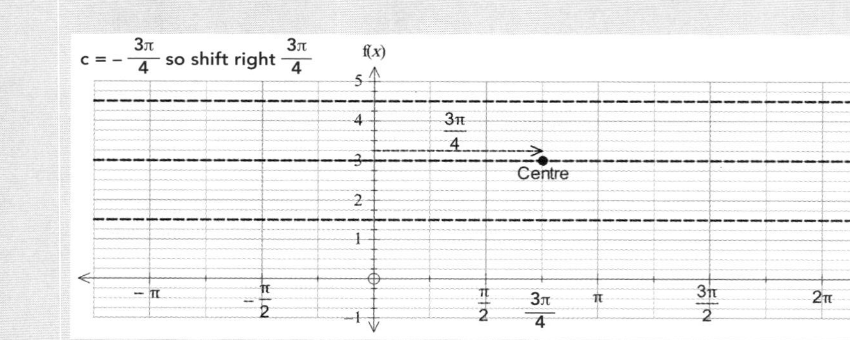

 ISBN: 9780170425728

Step 4: **b** tells you the **frequency**.

$$\mathbf{f(x) = 1.5\sin 2(x - \frac{3\pi}{4}) + 3}$$

b = 2

Calculate period using: period $= \frac{2\pi}{b} = \pi$.

Mark points either side of the centre at intervals of the period (•).

Make smaller marks halfway between these, (I).

Step 5: Use your knowledge of the shape of a sine graph to: **mark maximum** and **minimum** points between the points on the axis of symmetry.

Step 6: Use these lines and points to **complete** your graph.

Note: For a cosine function, the centre mark must be on the **maximum** line.

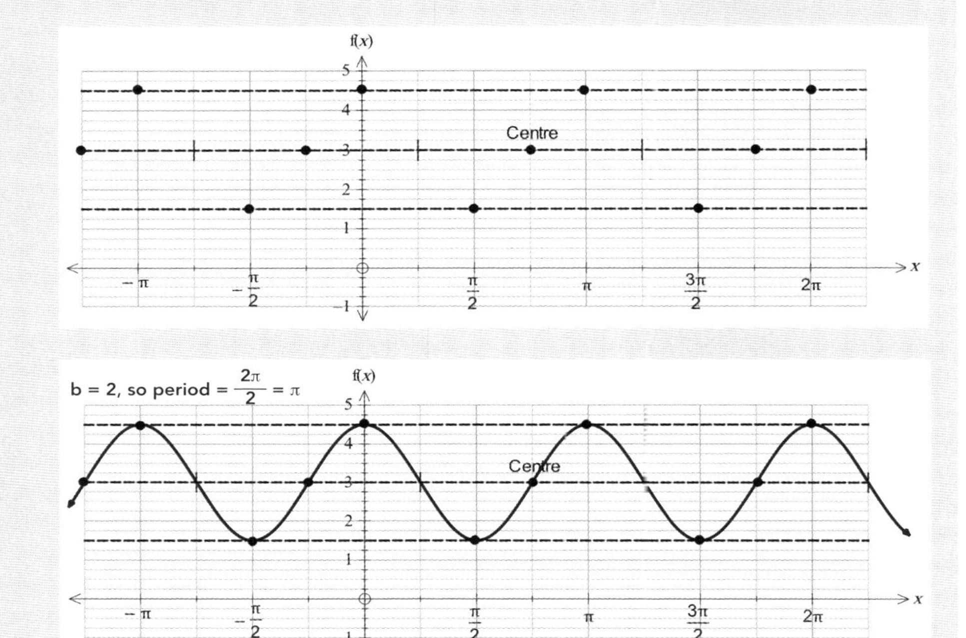

Match the following graphs to the equations below.

1

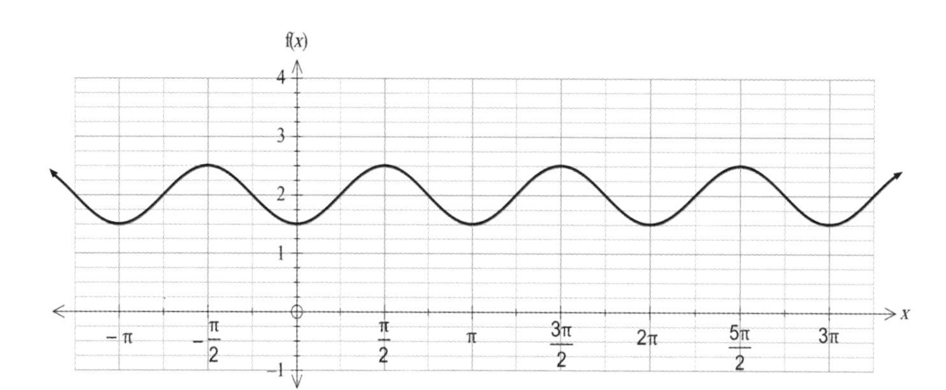

____

2

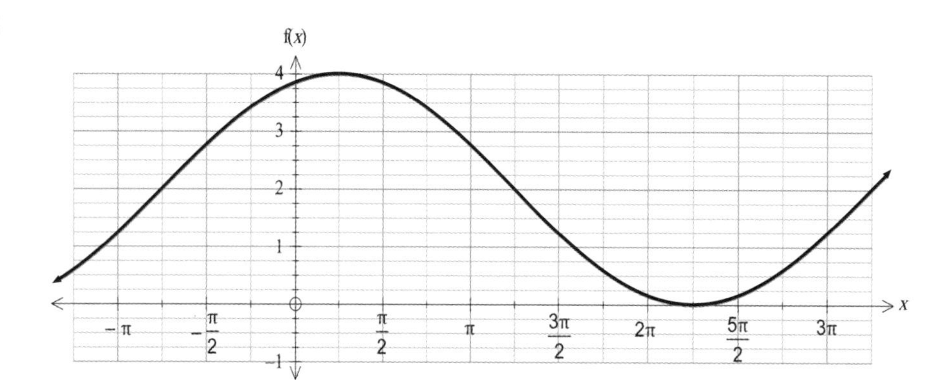

____

3

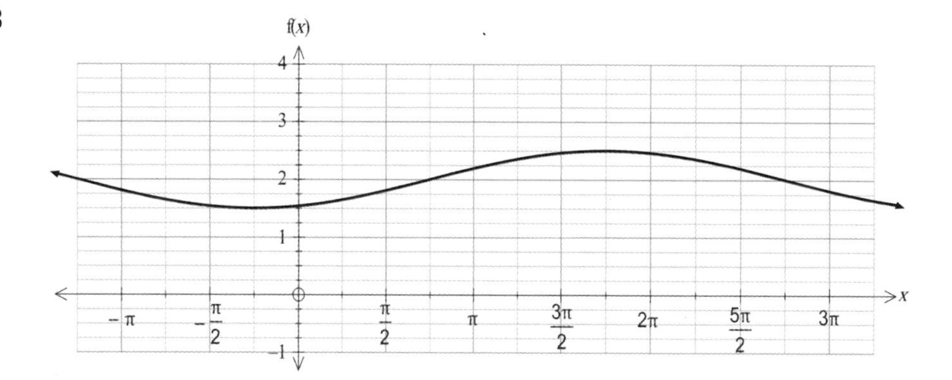

____

4

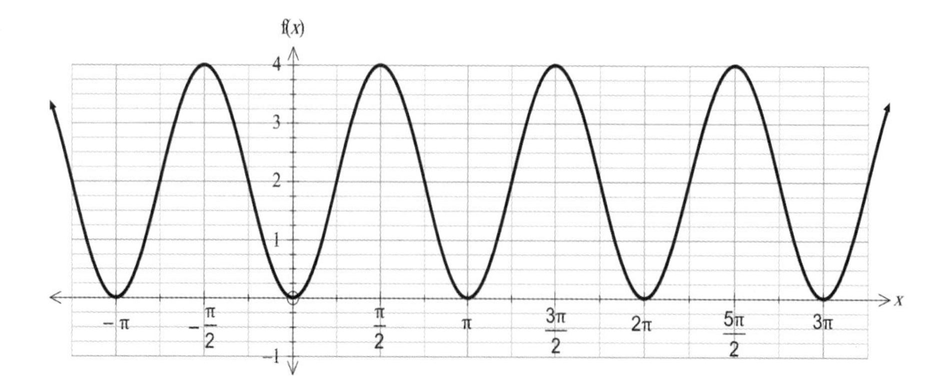

____

A $f(x) = 2\cos 2(x - \frac{\pi}{2}) + 2$

B $f(x) = 2\cos \frac{1}{2}(x - \frac{\pi}{4}) + 2$

C $f(x) = \frac{1}{2}\sin 2(x - \frac{\pi}{4}) + 2$

D $f(x) = \frac{1}{2}\sin 0.5(x - \frac{3\pi}{4}) + 2$

ISBN: 9780170425728

Write equations for the following graphs.

5 This graph is a function of f(x) = sin x.

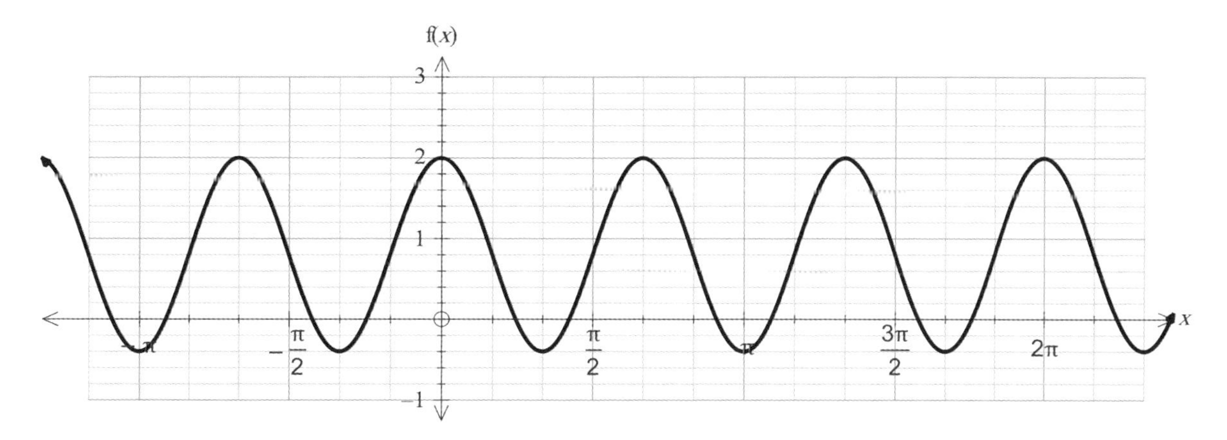

6 This graph is a function of f(x) = cos x.

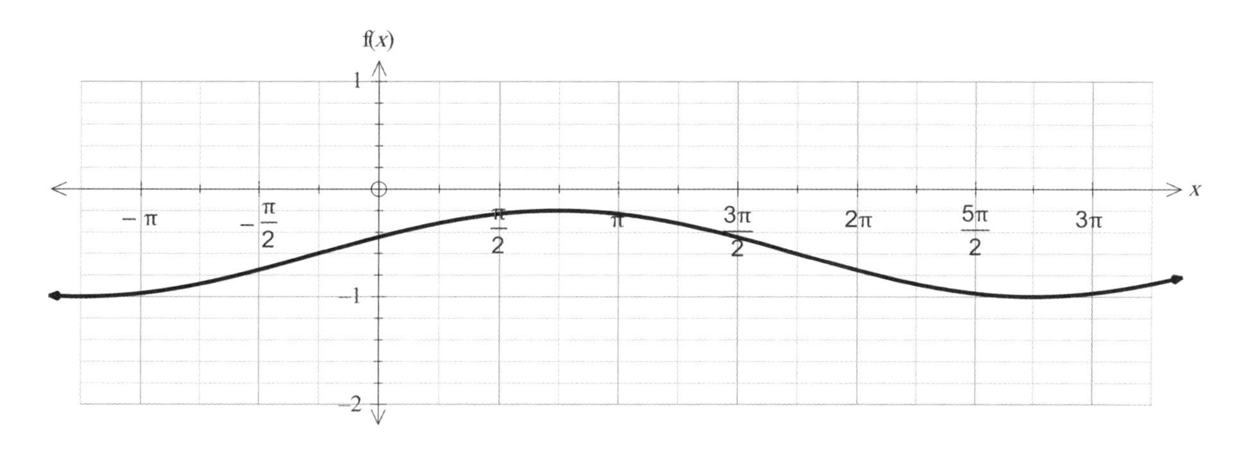

7 This graph is a function of f(x) = sin x.

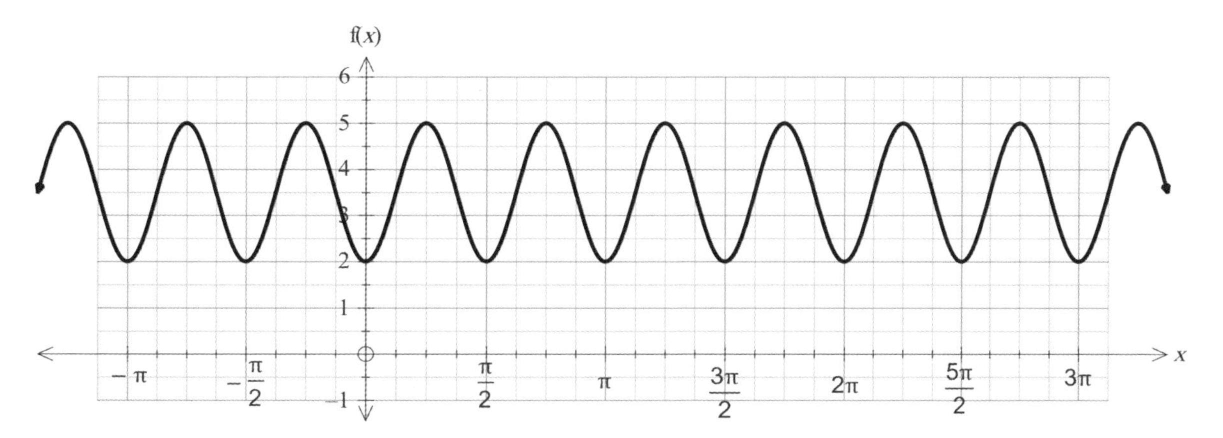

ISBN: 9780170425728

Draw graphs for the following equations.

8 $f(x) = 2\cos 0.5(x - \frac{\pi}{4}) + 3$

d = horizontal axis = ________

a = amplitude = ________, so maximum is at ________ and minimum is at ________.

c = horizontal translation = ________, so 'centre' is at (______, ______)

b = frequency = ________, so period = $\frac{2\pi}{\_\_\_}$ = ________.

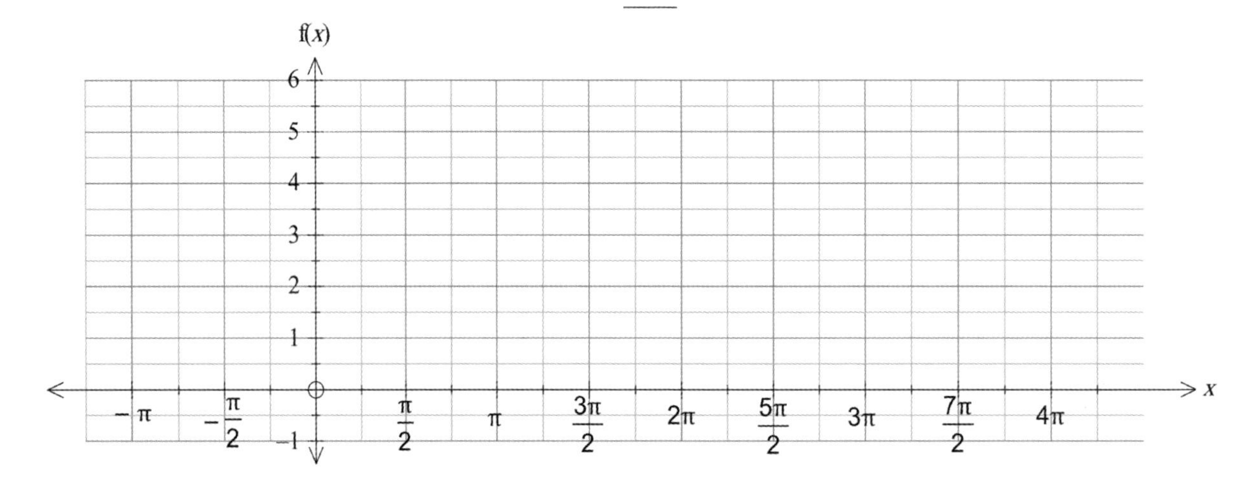

9 $f(x) = 0.8 \sin 3(x + \frac{\pi}{2}) - 2$

d = horizontal axis = ________

a = amplitude = ________, so maximum is at ________ and minimum is at ________.

c = horizontal translation = ________, so 'centre' is at (______, ______)

b = frequency = ________, so period = $\frac{2\pi}{\_\_\_}$ = ________.

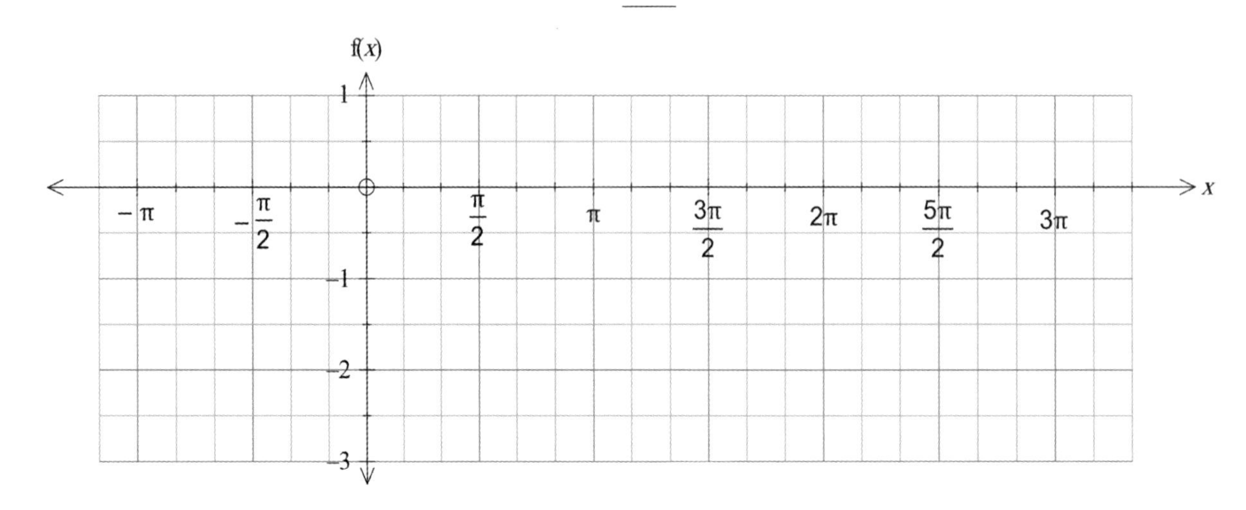

 ISBN: 9780170425728

10 $f(x) = 1.4\sin 0.8(x - \frac{3\pi}{8}) - 0.6$

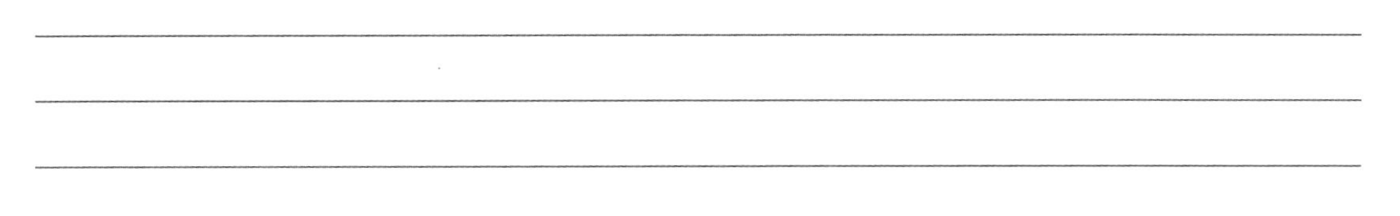

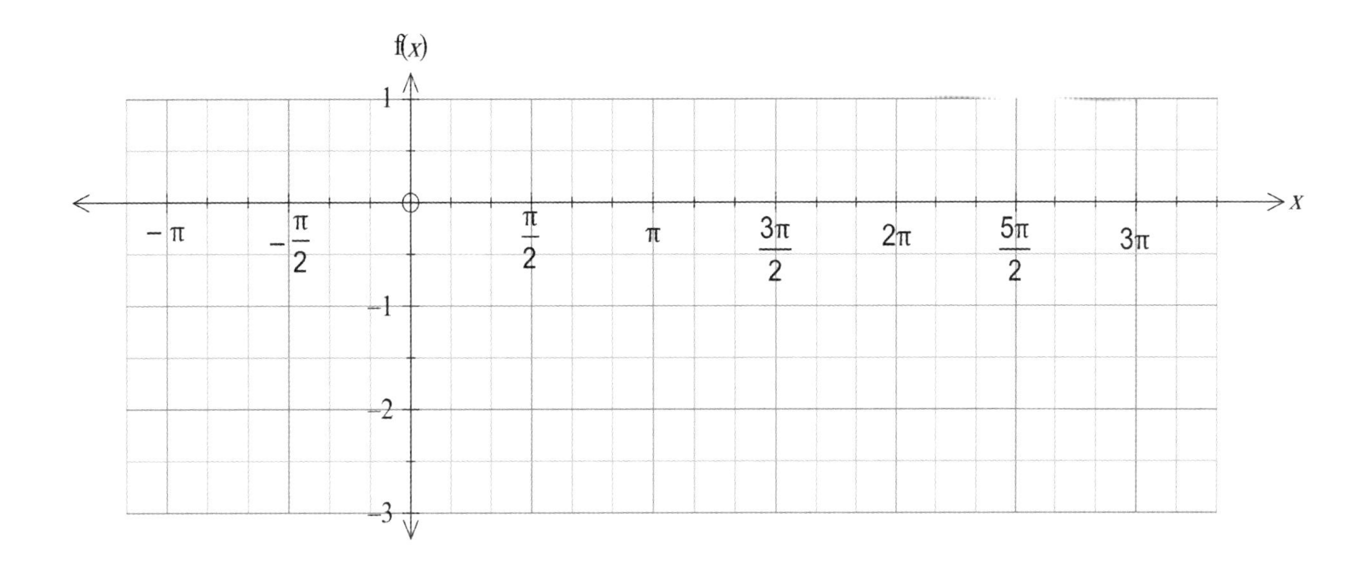

11 $f(x) = 3.5\cos 5(x + \frac{\pi}{5}) + 1.5$

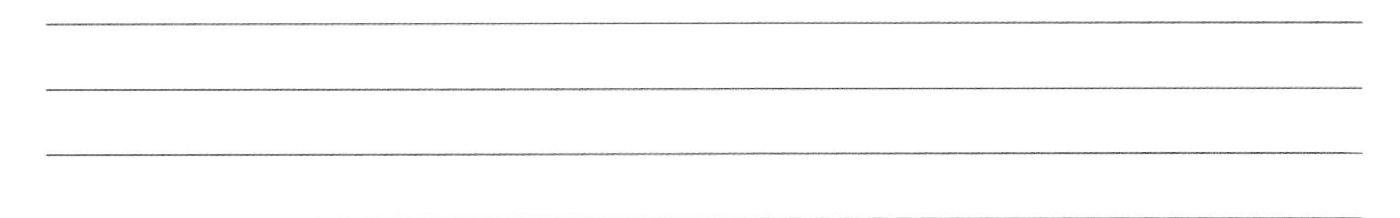

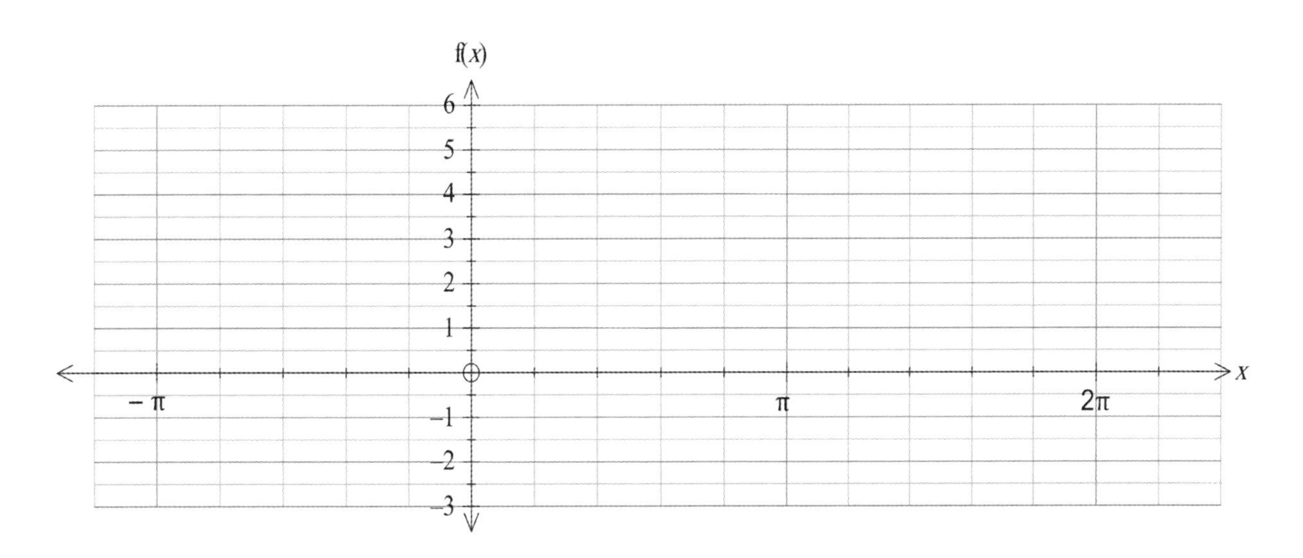

ISBN: 9780170425728

Using a graphics calculator to draw trigonometric graphs

Example: Draw the graph of $\mathbf{f(x) = 1.5\sin 2(x - \frac{3\pi}{4}) + 3}$

Check that your calculator is set to radians.

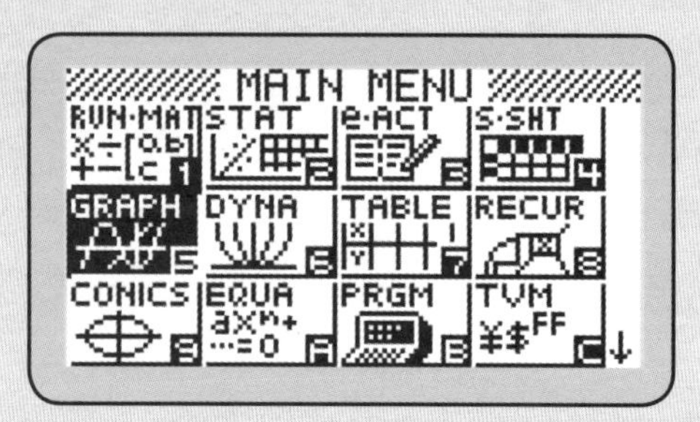

⇒ **MENU**
⇒ **GRAPH**
⇒ **EXE**
⇒ **Shift, SET UP**
⇒ ∇ ∇ ∇ ∇ ∇ ∇ ∇
⇒ **Angle**
⇒ **F2 Rad**
⇒ **EXE**

Go to Graph on the main menu.

MAIN MENU screen

⇒ **EXE**

Enter the function.

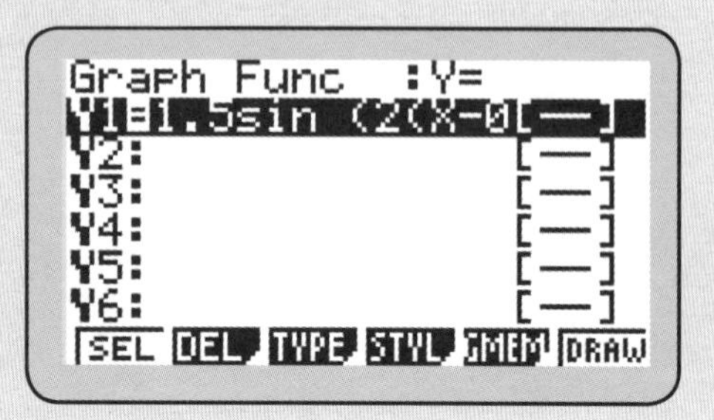

⇒ **1.5sin(2(x – 0.75π))+3**
⇒ **EXE**
⇒ **F6 Draw**

Note the extra pair of brackets.

ISBN: 9780170425728

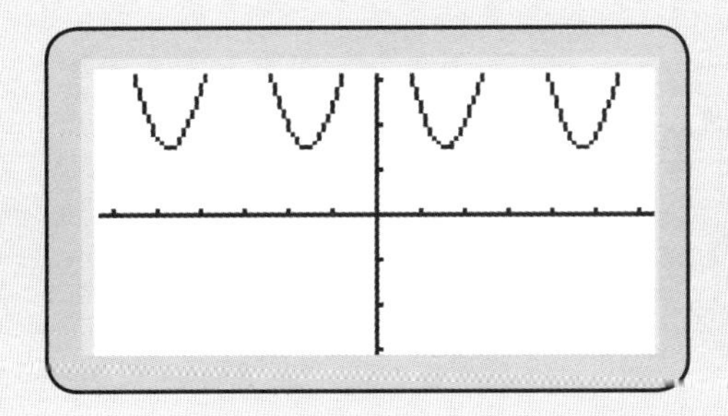

Often you will get a graph with an unsuitable range and/or domain. Also normal numbers on the *x*-axis are not as useful as radians.

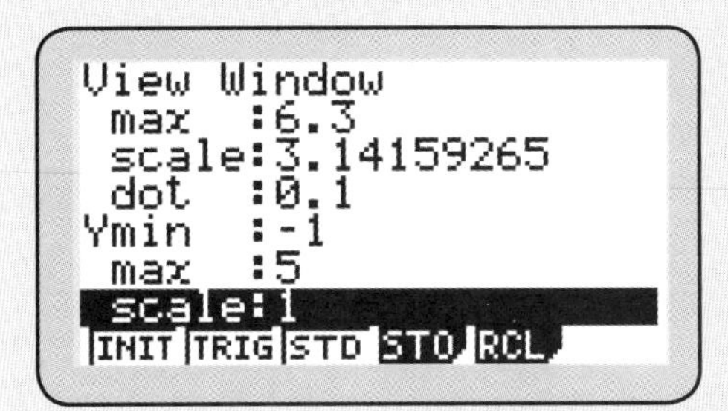

For this graph:

⇒ Shift F3 (V-window)

1 ∇ ∇ Change the scale to π. EXE

2 ∇ Change the Ymin to –1. EXE

3 Change the Ymax to 5. EXE

Note: You could also change the scale to 0.5 π, 0.333 π, etc.

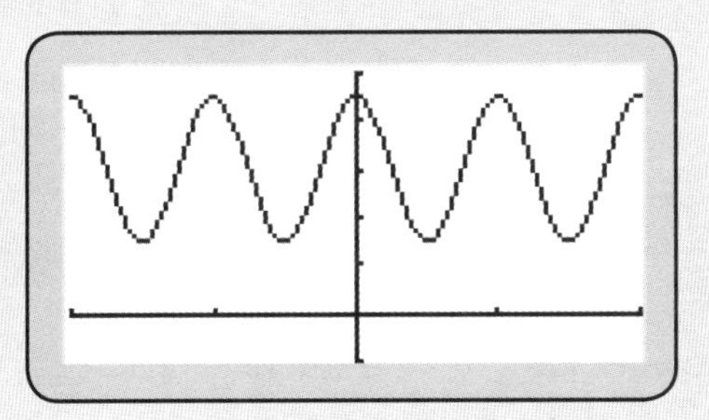

⇒ EXE

⇒ F6 Draw

This is a useful check for when you need to draw a function, but there are no gridlines, so its use at this stage is limited.

ISBN: 9780170425728

Trigonometric identities

- An **identity** is an equation that is true for **all** values of x.
- An **equation** is true for only **one or several** values of x, but not for all values.

Reciprocal trigonometric functions and identities

- You are familiar with sine, cosine and tangent functions.
- There are three new functions that you must become familiar with:

  cosec θ (**cosecant**)
  sec θ (**secant**)
  and cot θ (**cotangent**).

cosecant: y = cosec θ

In a right-angled triangle, **cosec** $\theta = \frac{\text{hypoteneuse}}{\text{opposite}} = \frac{\mathbf{1}}{\mathbf{sin}\ \theta}$

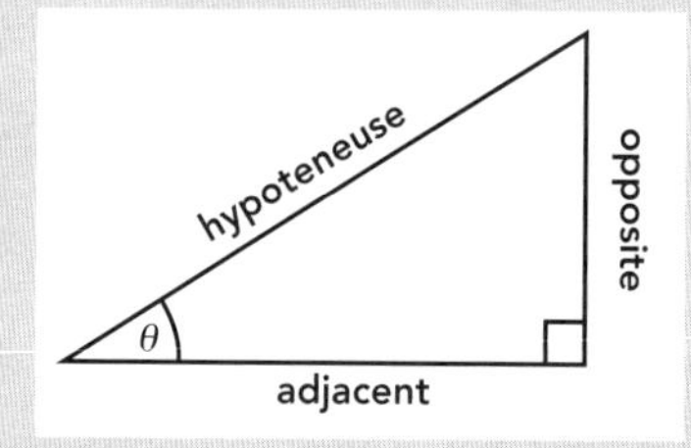

secant: y = sec θ

In a right-angled triangle, **sec** $\theta = \frac{\text{hypoteneuse}}{\text{adjacent}} = \frac{\mathbf{1}}{\mathbf{cos}\ \theta}$

cotangent: y = cot θ

In a right-angled triangle, **cot** $\theta = \frac{\text{adjacent}}{\text{opposite}} = \frac{\mathbf{1}}{\mathbf{tan}\ \theta}$

New trigonometric identities:

These are on your formula sheet.

cosec $\theta = \frac{1}{\sin\theta}$	**sec** $\theta = \frac{1}{\cos\theta}$	**cot** $\theta = \frac{1}{\tan\theta}$

- These are important because a common technique is to convert expressions so they contain only sines and cosines.
- Your calculator does not give values for these, so you need the identities in order to evaluate functions.

Examples:

1 Evaluate $\sec\frac{\pi}{3}$

$$\sec\frac{\pi}{3} = \frac{1}{\cos\frac{\pi}{3}} = \frac{1}{0.5} = 2$$

2 Evaluate $\text{cosec}^2\, 2.7 - 2\cot\frac{\pi}{4}$

$$\text{cosec}^2\, 2.7 - 2\cot\frac{\pi}{4} = \frac{1}{\sin^2 2.7} - 2 \times \frac{1}{\tan\frac{\pi}{4}}$$
$$= \frac{1}{(0.4274)^2} - 2 \times \frac{1}{1}$$
$$= 5.4743 - 2$$
$$= 3.4743$$

ISBN: 9780170425728

Evaluate the following.

1 $\text{cosec}\,\frac{\pi}{2}$

2 $\cot 2$

3 $\cot^2 \frac{\pi}{3}$

4 $4\text{cosec}\,5 - \sec^2 \pi$

5 $6\sec^2 \pi + \dfrac{1}{\cot \frac{\pi}{4}}$

6 $\cot^2 0.5 \times \dfrac{1}{\sec \pi} + 3\text{cosec}\,\dfrac{3\pi}{4}$

Identities and proofs

An identity = an equation which is always true for all values of x.

Structure:

Left Hand Side

Right Hand Side

A proof = a rigorous mathematical argument which demonstrates that an identity is true.

Structure:

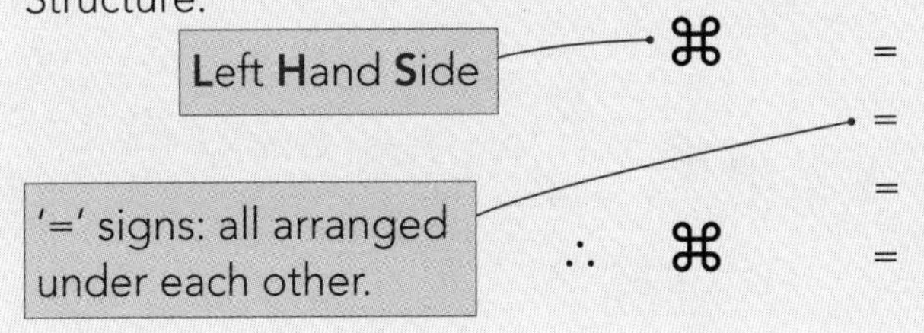

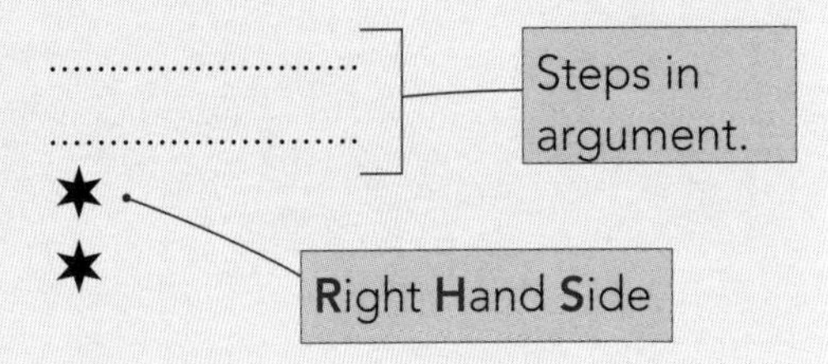

ISBN: 9780170425728

Hints for proving identities

1 Start with the more complex side.

2 If the expression contains squares, the Pythagorean identities are likely to be useful.

3 If one side of the expression has fractions, start with that side.

4 For an expression containing fractions which need to be added or subtracted, you will probably need to rewrite it with a common denominator:

$$\frac{a}{b} \pm \frac{c}{d} = \frac{ad \pm cb}{bd}$$

5 For an expression with a denominator which is a sum or difference, you will probably need to create the difference of two squares:

$$\frac{a}{b \pm c} = \frac{a}{b \pm c} \times \frac{b \mp c}{b \mp c}$$
$$= \frac{a(b \mp c)}{(b^2 - c^2)}$$

Then the denominator can often be converted to **one** term using a Pythagorean identity.

6 You will often have to deal with fractions divided by fractions. Remember:

$$\frac{\frac{a}{b}}{\frac{c}{d}} = \frac{a}{b} \times \frac{d}{c}$$

7 Look at the functions on the other side (the 'answer'): convert everything on the complex side to the same functions.

8 If you are stuck, convert everything to sines and cosines.

9 If you are still stuck, try starting from the other side.

10 You can sometimes reduce both sides to the same simple expression:

Left-hand side = simple expression
Right-hand side = the same simple expression

Note: There are sometimes different ways of completing a proof. If your answer is different, check it with your teacher.

ISBN: 9780170425728

Two more trigonometric identities and simple proofs

1 $\tan\theta = \dfrac{\sin\theta}{\cos\theta}$:

$$\sin\theta = \frac{O}{H} \text{ and } \cos\theta = \frac{A}{H}$$

$$\therefore \tan\theta = \frac{O}{A} = \frac{\frac{O}{H}}{\frac{A}{H}} = \frac{\sin\theta}{\cos\theta}$$

Divide numerator and denominator by H.

2 $\cot\theta = \dfrac{\cos\theta}{\sin\theta}$:

$$\cot\theta = \frac{1}{\tan\theta}$$

$$\therefore \cot\theta = \frac{\cos\theta}{\sin\theta}$$

New trigonometric identities:

$\tan\theta = \dfrac{\sin\theta}{\cos\theta}$	$\cot\theta = \dfrac{\cos\theta}{\sin\theta}$

We can combine these with the previous three identities:

$\operatorname{cosec}\theta = \dfrac{1}{\sin\theta}$	$\sec\theta = \dfrac{1}{\cos\theta}$	$\cot\theta = \dfrac{1}{\tan\theta}$

Note: These are all important for proofs involving identities.

Examples:

1 Prove that $\tan\theta \times \cos\theta = \sin\theta$

Convert everything to sin θ and cos θ.

$$\tan\theta \times \cos\theta = \frac{\sin\theta}{\cos\theta} \times \cos\theta = \sin\theta$$

When doing a proof, it is usually easiest to start with the more complicated side, and simplify it.

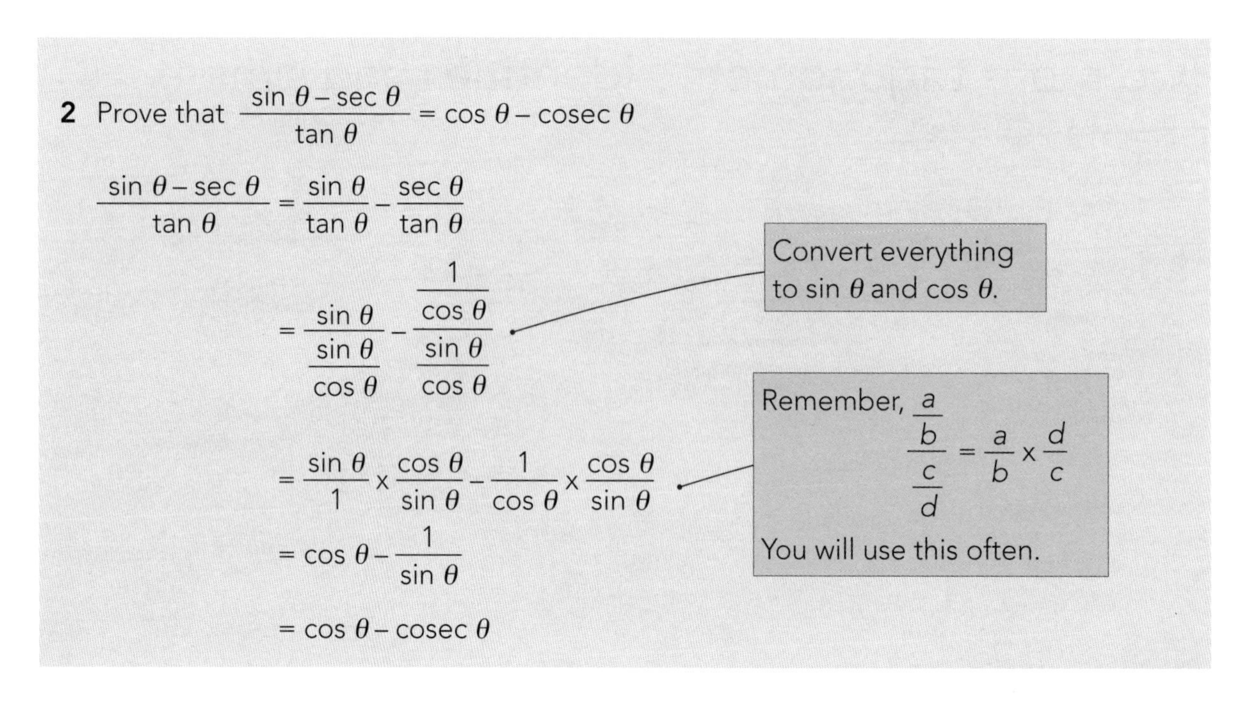

2 Prove that $\dfrac{\sin\theta - \sec\theta}{\tan\theta} = \cos\theta - \operatorname{cosec}\theta$

$$\frac{\sin\theta - \sec\theta}{\tan\theta} = \frac{\sin\theta}{\tan\theta} - \frac{\sec\theta}{\tan\theta}$$

$$= \frac{\sin\theta}{\frac{\sin\theta}{\cos\theta}} - \frac{\frac{1}{\cos\theta}}{\frac{\sin\theta}{\cos\theta}}$$

Convert everything to sin θ and cos θ.

$$= \frac{\sin\theta}{1} \times \frac{\cos\theta}{\sin\theta} - \frac{1}{\cos\theta} \times \frac{\cos\theta}{\sin\theta}$$

Remember, $\dfrac{\frac{a}{b}}{\frac{c}{d}} = \dfrac{a}{b} \times \dfrac{d}{c}$

You will use this often.

$$= \cos\theta - \frac{1}{\sin\theta}$$

$$= \cos\theta - \operatorname{cosec}\theta$$

Prove the following.

1 $\cot\theta \times \sin\theta = \cos\theta$

2 $\tan\theta \times \cos\theta \times \operatorname{cosec}\theta = 1$

3 $\dfrac{2\cos\theta}{\cot\theta} = 2\sin\theta$

 ISBN: 9780170425728

4 $\sec^2\theta \times \sin\theta \times \cot^2\theta = \operatorname{cosec}\theta$

5 $\dfrac{1+\cos\theta}{\sin\theta} = \operatorname{cosec}\theta + \cot\theta$

6 $\dfrac{\sin\theta + 3}{\cos\theta} = \tan\theta + 3\sec\theta$

7 $\dfrac{\sin\theta - \sec\theta}{\tan\theta} = \cos\theta - \operatorname{cosec}\theta$

ISBN: 9780170425728

8 $\tan\theta \times \cos^2\theta - \cot\theta \times \sin^2\theta = 0$

9 $\frac{3 - \tan\theta}{\sin\theta} - 2\text{cosec}\,\theta = \text{cosec}\,\theta - \sec\theta$

10 $\frac{3}{\cos\theta \times \text{cosec}\,\theta} - \frac{2\sin\theta \times \text{cosec}\,\theta}{\cot\theta} = \tan\theta$

ISBN: 9780170425728

The Pythagorean identities

$\sin\theta = \frac{O}{H} = \frac{Q}{1} = Q$

$\cos\theta = \frac{A}{H} = \frac{R}{1} = R$

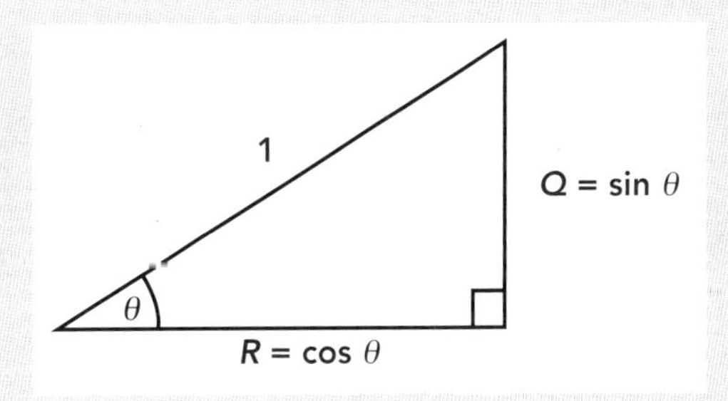

But by Pythagoras we know that $Q^2 + R^2 = 1$

$$\therefore \sin^2\theta + \cos^2\theta = 1$$

Divide this identity by $\cos^2\theta$:

$$\frac{\sin^2\theta}{\cos^2\theta} + \frac{\cos^2\theta}{\cos^2\theta} = \frac{1}{\cos^2\theta}$$

$$\therefore \tan^2\theta + 1 = \sec^2\theta$$

Divide this identity by $\sin^2\theta$:

$$\frac{\sin^2\theta}{\sin^2\theta} + \frac{\cos^2\theta}{\sin^2\theta} = \frac{1}{\sin^2\theta}$$

$$\therefore 1 + \cot^2\theta = \operatorname{cosec}^2\theta$$

New trigonometric identities:

These are on your formula sheet.

$\sin^2\theta + \cos^2\theta = 1$	$\tan^2\theta + 1 = \sec^2\theta$	$1 + \cot^2\theta = \operatorname{cosec}^2\theta$

Note: These are very useful when identities contain squares.

Examples:

1 $4\sin^2\theta - 3\cos^2\theta + 2\tan^2\theta = 7\sin^2\theta + 2\sec^2\theta - 5$

RHS contains $\sin^2\theta$, so don't change the first term.

$$\begin{aligned} 4\sin^2\theta - 3\cos^2\theta + 2\tan^2\theta &= 4\sin^2\theta - 3(1 - \sin^2\theta) + 2(\sec^2\theta - 1) \\ &= 4\sin^2\theta - 3 + 3\sin^2\theta + 2\sec^2\theta - 2 \\ &= 7\sin^2\theta + 2\sec^2\theta - 5 \end{aligned}$$

ISBN: 9780170425728

2 $\sec^4\theta - \tan^4\theta = 2\tan^2\theta + 1$

$$\sec^4\theta - \tan^4\theta = (\sec^2\theta - \tan^2\theta)(\sec^2\theta + \tan^2\theta)$$

Use the difference of two squares. $(a^2)^2 - (b^2)^2 = (a^2 - b^2)(a^2 + b^2)$

$$= ((\tan^2\theta + 1) - \tan^2\theta)((\tan^2\theta + 1) + \tan^2\theta)$$

$$= (1)(2\tan^2\theta + 1)$$

$$= 2\tan^2\theta + 1$$

3 $\dfrac{\sin A}{1-\cos A} = \dfrac{1+\cos A}{\sin A}$

A sum or difference (e.g. $1 - \cos A$) can often be eliminated by creating a difference of two squares.

$$\frac{\sin A}{1-\cos A} = \frac{\sin A}{1-\cos A} \times \frac{1+\cos A}{1+\cos A}$$

$$= \frac{\sin A(1+\cos A)}{1-\cos^2 A}$$

Then the difference of two squares can often be converted to a single expression using a Pythagorean identity.

$$= \frac{\sin A(1+\cos A)}{\sin^2 A}$$

$$= \frac{1+\cos A}{\sin A}$$

4 $\dfrac{2}{\sin\theta} = \dfrac{\sin\theta}{1-\cos\theta} + \dfrac{1-\cos\theta}{\sin\theta}$

The more complex side is on the right, so start with that.

$$\frac{\sin\theta}{1-\cos\theta} + \frac{1-\cos\theta}{\sin\theta} = \frac{\sin^2\theta + (1-\cos\theta)^2}{\sin\theta(1-\cos\theta)}$$

Where fractions with different denominators need to be added or subtracted, you must create a common denominator.

$$\frac{a}{b} \pm \frac{c}{d} = \frac{ad \pm cb}{bd}$$

You will use this often.

$$= \frac{\mathbf{sin^2}\,\theta + 1 - 2\cos\theta + \mathbf{cos^2}\,\theta}{\sin\theta(1-\cos\theta)}$$

$\sin^2\theta + \cos^2\theta = 1$

$$= \frac{2-2\cos\theta}{\sin\theta(1-\cos\theta)}$$

$$= \frac{2(1-\cos\theta)}{\sin\theta(1-\cos\theta)}$$

$$= \frac{2}{\sin\theta}$$

 ISBN: 9780170425728

Prove the following identities.

1 $\sin^2\theta + \tan^2\theta = \sec^2\theta - \cos^2\theta$

2 $\cos\theta(1 - \sin^2\theta) = \cos^3\theta$

3 $6\operatorname{cosec}^2\theta - 2(\cot^2\theta + 3) = 4\cot^2\theta$

4 $2\cot^2\theta = \operatorname{cosec}^4\theta - \cot^4\theta - 1$

5 $\dfrac{\cos\theta + \sin\theta}{\sin\theta} - 1 = \dfrac{\cos\theta}{\sin\theta}$

6 $\dfrac{\cos A}{1 + \sin A} = \dfrac{1 - \sin A}{\cos A}$

7 $\dfrac{\cos\theta}{1 + \sin\theta} - \dfrac{1 - \sin\theta}{\cos\theta} = 0$

ISBN: 9780170425728

Mixing it up

Once you combine the use of many identities, it is important that you state your **reasoning** for most lines of working.

Combining all of these:

$\text{cosec}\ \theta = \frac{1}{\sin\theta}$	$\sec\theta = \frac{1}{\cos\theta}$	$\cot\theta = \frac{1}{\tan\theta}$

$\tan\theta = \frac{\sin\theta}{\cos\theta}$	$\cot\theta = \frac{\cos\theta}{\sin\theta}$

$\sin^2\theta + \cos^2\theta = 1$	$\tan^2\theta + 1 = \sec^2\theta$	$1 + \cot^2\theta = \text{cosec}^2\,\theta$

Examples:

1 $\cot\theta + \tan\theta = \sec\theta\,\text{cosec}\,\theta$

$$\cot\theta + \tan\theta = \frac{\cos\theta}{\sin\theta} + \frac{\sin\theta}{\cos\theta}$$

$$= \frac{\cos^2\theta + \sin^2\theta}{\sin\theta\cos\theta}$$

$$= \frac{1}{\sin\theta\cos\theta}$$

$$= \sec\theta\,\text{cosec}\,\theta$$

2 $\frac{\sin^4\theta - \cos^4\theta}{\sin^2\theta\cos^2\theta} = \sec^2\theta - \text{cosec}^2\,\theta$

$$\frac{\sin^4\theta - \cos^4\theta}{\sin^2\theta\cos^2\theta} = \frac{(\sin^2\theta - \cos^2\theta)(\sin^2\theta + \cos^2\theta)}{\sin^2\theta\cos^2\theta}$$

$$= \frac{(\sin^2\theta - \cos^2\theta)(1)}{\sin^2\theta\cos^2\theta}$$

$$= \frac{\sin^2\theta}{\sin^2\theta\cos^2\theta} - \frac{\cos^2\theta}{\sin^2\theta\cos^2\theta}$$

$$= \frac{1}{\cos^2\theta} - \frac{1}{\sin^2\theta}$$

$$= \sec^2\theta - \text{cosec}^2\,\theta$$

Remember:
$$\frac{a-b}{cd} = \frac{a}{cd} - \frac{b}{cd}$$
This converts a single fraction into a difference (or sum).

3 $(\sec\theta - \cos\theta)(\csc\theta - \sin\theta) = \dfrac{1}{\tan\theta + \cot\theta}$

$$(\sec\theta - \cos\theta)(\operatorname{cosec}\theta - \sin\theta) = \left(\frac{1}{\cos\theta} - \cos\theta\right)\left(\frac{1}{\sin\theta} - \sin\theta\right)$$

$$= \frac{1 - \cos^2\theta}{\cos\theta} \times \frac{1 - \sin^2\theta}{\sin\theta}$$

$$= \frac{\sin^2\theta}{\cos\theta} \times \frac{\cos^2\theta}{\sin\theta}$$

$$= \sin\theta\cos\theta$$

This is a very simple expression involving just sine and cosine. When this happens, it is often easiest to try to simplify the **other** side of the identity to the same expression.

$$\frac{1}{\tan\theta + \cot\theta} = \frac{1}{\dfrac{\sin\theta}{\cos\theta} + \dfrac{\cos\theta}{\sin\theta}}$$

$$= \frac{1}{\dfrac{\sin^2\theta + \cos^2\theta}{\cos\theta\sin\theta}}$$

$$= \frac{1}{\dfrac{1}{\cos\theta\sin\theta}}$$

$$= \cos\theta\sin\theta$$

Prove the following identities.

1 $\dfrac{1}{1 + \sin\theta} + \dfrac{1}{1 - \sin\theta} = 2\sec^2\theta$

2 $(\cot^2\theta + 1)(\sec^2\theta - 1) = \sec^2\theta$

 ISBN: 9780170425728

3 $(1 - \sin\theta)(1 + \operatorname{cosec}\theta) = \cos\theta\cot\theta$

4 $\dfrac{\tan^2\theta + 1}{\sec\theta} = \dfrac{1 - \sin^2\theta}{\cos^3\theta}$

5 $\dfrac{\cos^4\theta - \sin^4\theta}{\sin^2\theta} = \cot^2\theta - 1$

6 $\dfrac{1 + \sin\theta}{\cos\theta} = \dfrac{\cos\theta}{1 - \sin\theta}$

ISBN: 9780170425728

7 $\cos\theta - \cot^2\theta \times \sec\theta = \sin\theta\tan\theta$

8 $\dfrac{\sec\theta}{\tan\theta - 1} - \dfrac{\tan\theta - 1}{\sec\theta} = \dfrac{2\sin\theta}{\tan\theta - 1}$

9 $\dfrac{\sec\theta}{\sec\theta - \cos\theta} = \text{cosec}^2\,\theta$

10 $\dfrac{1 - \sin\theta}{\cos^3\theta} = \dfrac{\sec\theta}{\sin\theta + 1}$

ISBN: 9780170425728

Angle sums and differences

Unlike most mathematical situations, in trigonometry:

$\tan(A + B) \neq \tan A + \tan B$ and $\sin(A - B) \neq \sin A - \sin B$, etc.

These are the new identities for finding a trigonometric function of a sum or difference of two angles:

$\sin(\boldsymbol{A} \pm \boldsymbol{B}) = \sin \boldsymbol{A} \cos \boldsymbol{B} \pm \cos \boldsymbol{A} \sin \boldsymbol{B}$
$\cos(\boldsymbol{A} \pm \boldsymbol{B}) = \cos \boldsymbol{A} \cos \boldsymbol{B} \mp \sin \boldsymbol{A} \sin \boldsymbol{B}$
$\tan(\boldsymbol{A} \pm \boldsymbol{B}) = \dfrac{\tan \boldsymbol{A} \pm \tan \boldsymbol{B}}{1 \mp \tan \boldsymbol{A} \tan \boldsymbol{B}}$

1 Simplifying expressions containing angle sums and differences

- Using your knowledge of sine, cosine and tangent functions, it is often possible to simplify expressions containing sums and differences of angles.
- You may find it useful to have a sketch of sine and cosine functions up to 2π to refer to:

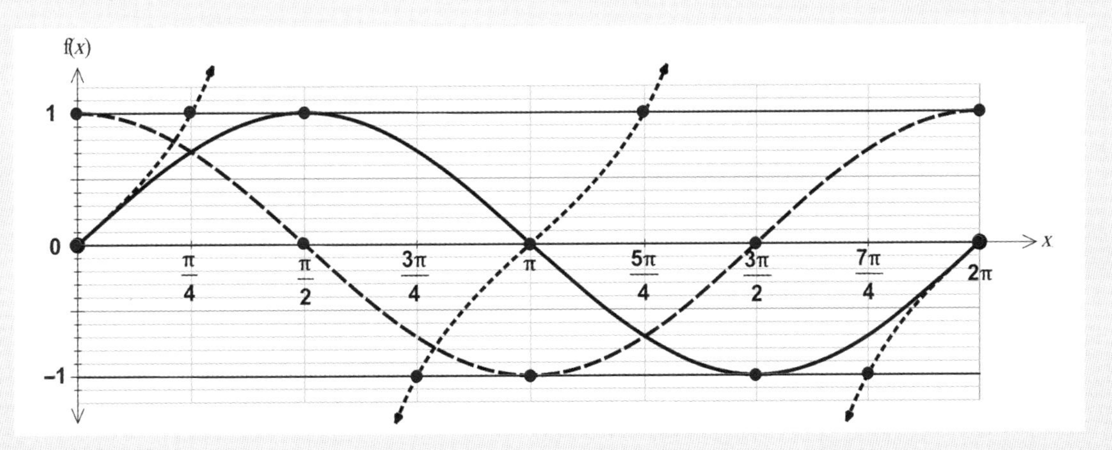

Examples: Simplify the following.

1 $\sin(a + \frac{3\pi}{2})$

$$\sin(a + \frac{3\pi}{2}) = \sin a \times \cos \frac{3\pi}{2} + \cos a \times \sin \frac{3\pi}{2}$$
$$= \sin a \times (0) + \cos a \times (-1)$$
$$= -\cos a$$

2 $\cot(b - \frac{7\pi}{4})$

$$\cot\left(b - \frac{7\pi}{4}\right) = \frac{1 + \tan b \times \tan \frac{7\pi}{4}}{\tan b - \tan \frac{7\pi}{4}}$$
$$= \frac{1 + \tan b \times (-1)}{\tan b - (-1)}$$
$$= \frac{1 - \tan b}{\tan b + 1}$$

Remember: $\cot \theta = \dfrac{1}{\tan \theta}$

ISBN: 9780170425728

Simplify the following expressions.

1 $\sin(2\pi + A)$

2 $\cos(a + \frac{\pi}{2})$

3 $\tan(b - \pi)$

4 $\sec(\frac{\pi}{2} - A)$

5 $\operatorname{cosec}(b + \frac{3\pi}{2})$

6 

$\cot(B + 4\pi)$

ISBN: 9780170425728

2 Finding exact values of angle sums and differences

- You already know how to find exact values for trigonometric functions of $\frac{\pi}{4}$, $\frac{\pi}{3}$ and $\frac{\pi}{6}$.

From page 12, recall that:

$\sin 45° = \sin \frac{\pi}{4} = \frac{1}{\sqrt{2}}$	$\cos 45° = \cos \frac{\pi}{4} = \frac{1}{\sqrt{2}}$	$\tan 45° = \tan \frac{\pi}{4} = 1$
$\sin 60° = \sin \frac{\pi}{3} = \frac{\sqrt{3}}{2}$	$\cos 60° = \cos \frac{\pi}{3} = \frac{1}{2}$	$\tan 60° = \tan \frac{\pi}{3} = \sqrt{3}$
$\sin 30° = \sin \frac{\pi}{6} = \frac{1}{2}$	$\cos 30° = \cos \frac{\pi}{6} = \frac{\sqrt{3}}{2}$	$\tan 30° = \tan \frac{\pi}{6} = \frac{1}{\sqrt{3}}$

- You can now use the formulas for finding the sums and differences of angles to find exact values for trigonometric functions of many other angles.
- Do **not** convert your answer to a decimal unless it terminates (e.g. $\frac{1}{2} = 0.5$) or is recurring (e.g. $\frac{1}{11} = 0.\dot{0}\dot{9}$). Any other decimal will be rounded, so not exact.

Examples: Find exact values for the following.

1 $\sin 210°$

$$\sin 210° = \sin (180° + 30°)$$
$$= \sin 180° \times \cos 30° + \cos 180° \times \sin 30°$$
$$= 0 \times \frac{\sqrt{3}}{2} + (-1) \times \frac{1}{2}$$
$$= -\frac{1}{2}$$

2 $\cot \frac{4\pi}{3}$

$$\cot \frac{4\pi}{3} = \frac{1}{\tan\left(\pi + \frac{\pi}{3}\right)}$$
$$= \frac{1 - \tan \pi \times \tan \frac{\pi}{3}}{\tan \pi + \tan \frac{\pi}{3}}$$
$$= \frac{1 - 0 \times \sqrt{3}}{0 + \sqrt{3}}$$
$$= \frac{1}{\sqrt{3}}$$

Leave your answer in this form. Do not convert it to a decimal.

ISBN: 9780170425728

Find exact values for the following.

1 $\cos 15°$ Hint: $\sqrt{2} \times \sqrt{2} = 2$.

2 $\tan 105°$

3 $\sin \frac{5\pi}{12}$

4 $\sec \frac{7\pi}{12}$

5 $\operatorname{cosec} 225°$

6 $\sec \frac{11\pi}{6}$

ISBN: 9780170425728

3 Proving identities with expressions containing angle sums and differences

- You can use similar methods to those you have used earlier.

Prove the following identities.

1 $\cos(A + B) - \cos(A - B) = -2\sin A \sin B$

2 $\sin(A + B)\sin(A - B) = \sin^2 A - \sin^2 B$

3 $\dfrac{\sin(x - y)}{\cos x \cos y} = \tan x - \tan y$

4 $\sin\left(\dfrac{\pi}{4} + A\right) + \sin\left(\dfrac{\pi}{4} - A\right) = \sqrt{2}\cos A$

ISBN: 9780170425728

5 $\dfrac{\cos\left(\frac{\pi}{4} - \theta\right)}{\cos\frac{\pi}{4}\cos\theta} - \tan\theta = 1$

Prove the following identities and then use the graphs to show that each is true.

6 $\tan(\pi + \theta) = \tan\theta$

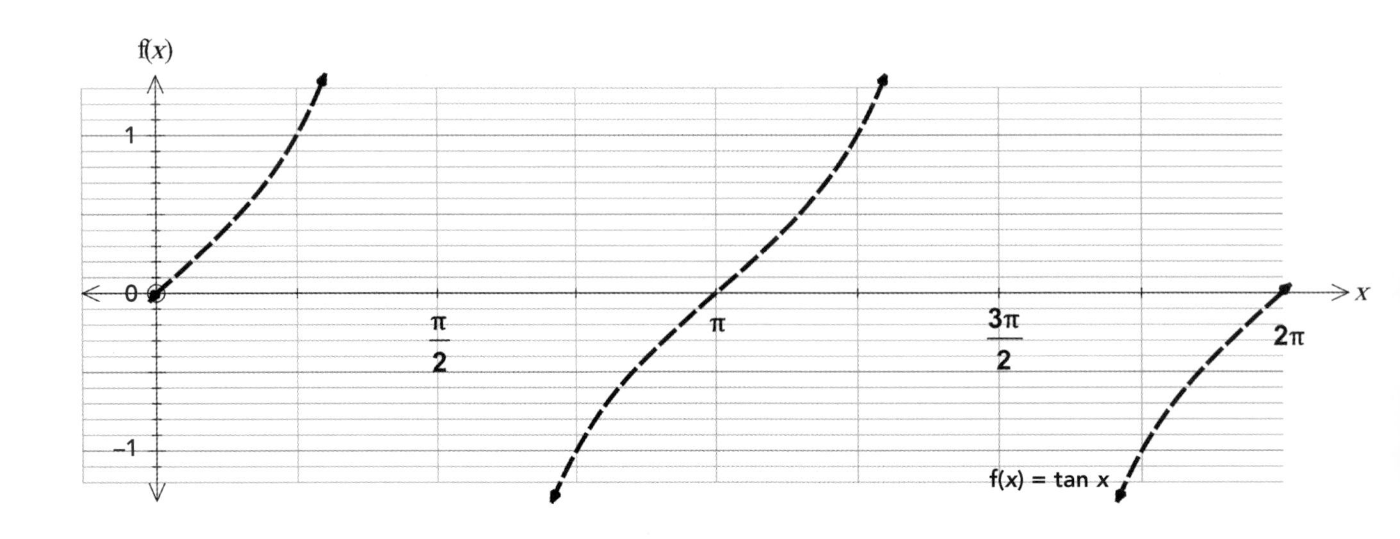

 ISBN: 9780170425728

7 $\sin(\pi - \theta) = \sin\theta$

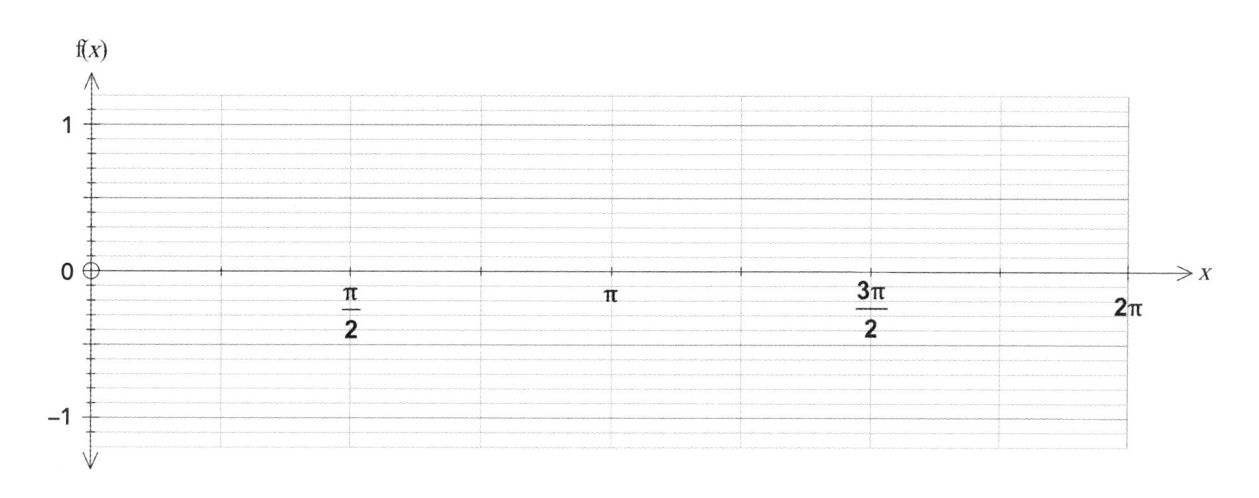

8 $\cos\left(\frac{\pi}{2} + \theta\right) = -\sin\theta$

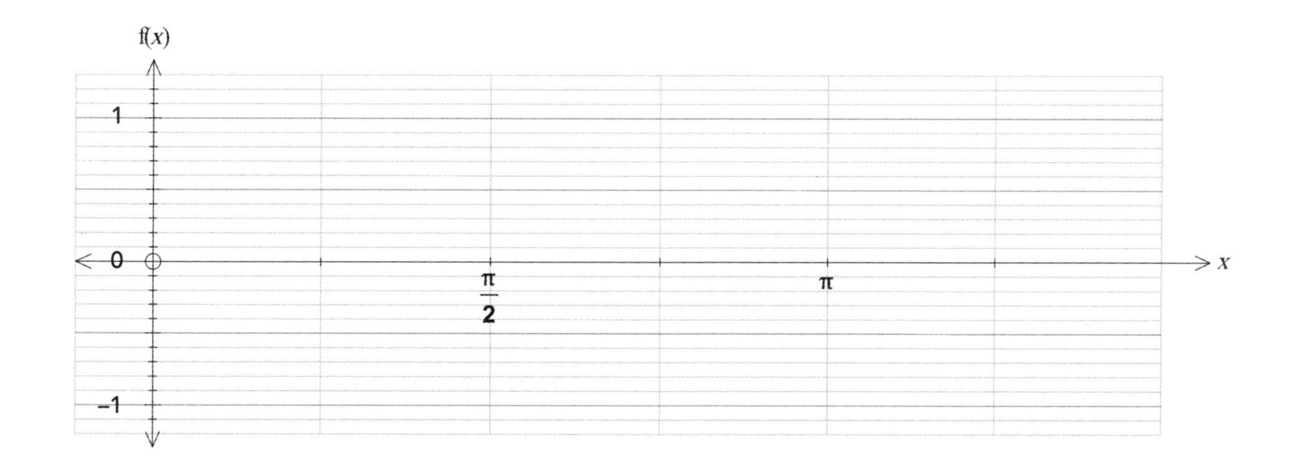

ISBN: 9780170425728

4 Bits and pieces

1 Find the exact value of $\dfrac{\tan 48° + \tan 12°}{1 - \tan 48° \tan 12°}$

2 Simplify $\sin\left(A - \frac{\pi}{6}\right) + \cos\left(A + \frac{\pi}{3}\right)$

3 If $\tan A = \frac{1}{2}$ and $\tan B = \frac{1}{4}$, find the value of $\tan (A + B)$.

4 If $5\cos\left(A + \frac{\pi}{4}\right) = \sin\left(A - \frac{\pi}{4}\right)$, find the value of tan A.

ISBN: 9780170425728

Double angles

- Unlike most mathematical situations, in trigonometry:

$$\sin 2A \neq 2\sin A, \qquad \cos 2A \neq 2\cos A \qquad \text{and} \qquad \tan 2A \neq 2\tan A$$

- The formulae for calculating sums and differences of angles can be used to find the correct formulae for double angles.

Sines of double angles:

$$\begin{aligned}\sin 2A &= \sin A\cos A + \cos A\sin A\\ &= 2\sin A\cos A\end{aligned}$$

Cosines of double angles:

$$\begin{aligned}\cos 2A &= \cos A\cos A - \sin A\sin A\\ &= \cos^2 A - \sin^2 A\\ \text{or} \quad &= (1-\sin^2 A) - \sin^2 A\\ &= 1 - 2\sin^2 A\\ \text{or} \quad &= \cos^2 A - (1-\cos^2 A)\\ &= 2\cos^2 A - 1\end{aligned}$$

Substitute $\cos^2 A = 1 - \sin^2 A$

Substitute $\sin^2 A = 1 - \cos^2 A$

Tangents of double angles:

$$\begin{aligned}\tan 2A &= \frac{\tan A + \tan A}{1 - \tan A\tan A}\\ &= \frac{2\tan A}{1-\tan^2 A}\end{aligned}$$

These are the new identities for finding a trigonometric function of a double angle:

$\sin 2A = 2\sin A\cos A$
$\cos 2A = \cos^2 A - \sin^2 A = 2\cos^2 A - 1 = 1 - 2\sin^2 A$
$\tan 2A = \dfrac{2\tan A}{1-\tan^2 A}$

You will often need to think carefully about which of these you use.

ISBN: 9780170425728

1 Using the formulae

It is useful to recognise these formulae.

Examples: Write each expression as a single trigonometric ratio.

1 $1 - 2\sin^2 \frac{\pi}{4} = \cos\left(2 \times \frac{\pi}{4}\right)$

$= \cos \frac{\pi}{2}$

2 $\frac{2\tan 4\theta}{1 - \tan^2 4\theta} = \tan(2 \times 4\theta)$

$= \tan 8\theta$

Write the following expressions as single trigonometric ratios.

1 $2\sin 60° \cos 60°$

2 $\cos^2 \frac{\pi}{2} - \sin^2 \frac{\pi}{2}$

3 $\frac{2\tan A}{1 - \tan^2 A}$

4 $1 - 2\cos^2 \frac{\pi}{8}$

5 $\cos 3 \sin 3$

6 $\sin^2 \frac{\pi}{3} - \cos^2 \frac{\pi}{3}$

ISBN: 9780170425728

2 Finding exact values of double angles

Examples: Find exact values for the following.

1 $\cos 2\theta$ if $0 < \theta < \frac{\pi}{2}$ and $\cos\theta = 0.5$

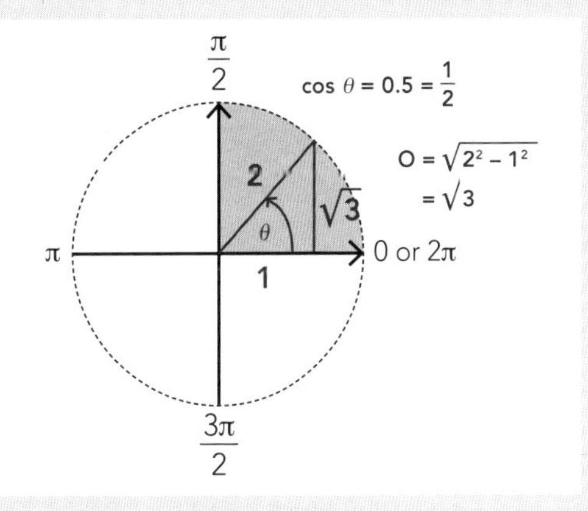

$$\cos 2\theta = \cos^2\theta - \sin^2\theta$$

$$= \left(\frac{1}{2}\right)^2 - \left(\frac{\sqrt{3}}{2}\right)^2$$

$$= \frac{1}{4} - \frac{3}{4}$$

$$= -\frac{1}{2}$$

Note the negative answer. While θ is in the first quadrant, 2θ is in the second quadrant where cosines are negative.

2 $\cot 2\theta$ if $\frac{\pi}{2} < \theta < \pi$ and $\sin\theta = \frac{1}{3}$

$$\cot 2\theta = \frac{1 - \tan^2\theta}{2\tan\theta}$$

$$= \frac{1 - \left(\frac{-1}{\sqrt{8}}\right)^2}{2 \times \frac{1}{-\sqrt{8}}}$$

$$= \frac{1 - \frac{1}{8}}{\frac{-2}{\sqrt{8}}}$$

$$= -\frac{7}{8} \times \frac{\sqrt{8}}{2}$$

$$= -\frac{7}{8} \times \frac{2\sqrt{2}}{2}$$

$$= -\frac{7\sqrt{2}}{8}$$

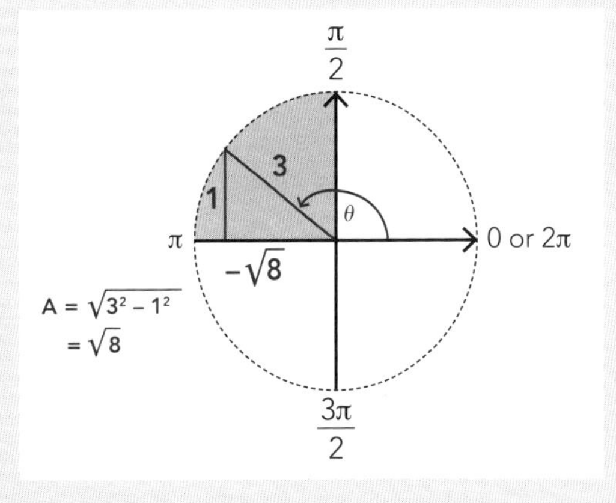

Answer the following.

1 If $\tan\theta = \frac{1}{3}$ and $0 < \theta < \pi$, evaluate each of the following.

a $\sin 2\theta$

b $\cos 2\theta$

c $\tan 2\theta$

2 If $\tan\theta = -\sqrt{3}$ and $\frac{\pi}{2} < \theta < \pi$, evaluate each of the following.

a $\sin 2\theta$

b $\cos 2\theta$

c $\tan 2\theta$

 ISBN: 9780170425728

3 If $\sin 2\theta = \frac{4}{5}$ and $0 < \theta < \frac{\pi}{5}$, evaluate each of the following.

a $\sin 4\theta$

b $\cos 4\theta$

c $\tan 4\theta$

4 If $\tan \theta = 3$ and $\pi < \theta < \frac{3\pi}{2}$, evaluate each of the following.

a $\sin 2\theta$

b $\cos 2\theta$

c $\tan 2\theta$

ISBN: 9780170425728

3 Using the formula to find half angles

- When given the value of a trigonometric function of θ, the double angle formulas can be used to find trigonometric functions of $\frac{\theta}{2}$.

Examples:

1 If $\cos = \frac{4}{5}$ and θ is in the first quadrant, then find the exact value of $\sin\frac{\theta}{2}$.

$$\cos\theta = \frac{4}{5} = 1 - 2\sin^2\frac{\theta}{2}$$

$$\therefore 2\sin^2\frac{\theta}{2} = 1 - \frac{4}{5}$$

$$\sin^2\frac{\theta}{2} = \frac{1}{10}$$

$$\sin\frac{\theta}{2} = \pm\frac{1}{\sqrt{10}}$$

Note that the only way to find $\sin\frac{\theta}{2}$ is to use $\cos\theta = 1 - 2\sin^2\frac{\theta}{2}$.

θ is in the first quadrant, so $\frac{\theta}{2}$ must also be in the first quadrant

$$\therefore \sin\frac{\theta}{2} = \frac{1}{\sqrt{10}}$$

2 Find the exact value of $\cos\frac{\pi}{8}$.

$$\cos\frac{\pi}{4} = \frac{1}{\sqrt{2}} = 2\cos^2\frac{\pi}{8} - 1$$

$$2\cos^2\frac{\pi}{8} = 1 + \frac{1}{\sqrt{2}} = \frac{\sqrt{2}+1}{\sqrt{2}}$$

$$\cos^2\frac{\pi}{8} = \frac{\sqrt{2}+1}{2\sqrt{2}} \times \frac{\sqrt{2}}{\sqrt{2}}$$

$$= \frac{2+\sqrt{2}}{4}$$

$$\cos\frac{\pi}{8} = \pm\sqrt{\frac{2+\sqrt{2}}{4}}$$

$$= \pm\frac{\sqrt{2+\sqrt{2}}}{2}$$

Note that the only way to find $\cos\frac{\theta}{2}$ is to use $\cos\theta = 2\cos^2\frac{\theta}{2} - 1$.

Multiplying by $\frac{\sqrt{2}}{\sqrt{2}}$ means there is no longer a square root in the denominator.

$\cos\frac{\pi}{4}$ is in the first quadrant, so $\cos\frac{\pi}{8}$ must also be in the first quadrant

$$\therefore \cos\frac{\pi}{8} = \frac{\sqrt{2+\sqrt{2}}}{2}$$

 ISBN: 9780170425728

3 If $\sin\theta = -\frac{5}{13}$ and $\pi < \theta < \frac{3\pi}{2}$, find the value of $\tan\frac{\theta}{2}$.

$\sin\theta = -\frac{5}{13} \Rightarrow \tan\theta = \frac{5}{12}$

$$\tan\theta = \frac{5}{12} = \frac{2\tan\frac{\theta}{2}}{1-\tan^2\frac{\theta}{2}}$$

Note that the only way to find $\tan\frac{\theta}{2}$ is to use $\tan\theta = \frac{2\tan\frac{\theta}{2}}{1-\tan^2\frac{\theta}{2}}$ and this will require the solving of a quadratic equation. This one can be factorised, but you will need to use your calculator to solve others.

$$\therefore \quad 5 - 5\tan^2\frac{\theta}{2} = 24\tan\frac{\theta}{2}$$

$$5\tan^2\frac{\theta}{2} + 24\tan\frac{\theta}{2} - 5 = 0$$

$$\left(5\tan\frac{\theta}{2} - 1\right)\left(\tan\frac{\theta}{2} + 5\right) = 0$$

$$\therefore \quad \tan\frac{\theta}{2} = \frac{1}{5} \text{ or } \tan\frac{\theta}{2} = -5$$

If θ is in the third quadrant, then $\frac{\theta}{2}$ must be in the second quadrant, where tangents are negative.

$$\therefore \quad \tan\frac{\theta}{2} = -5$$

Answer the following questions.

1 If $\sin\theta = \frac{4}{5}$, find $\cos\frac{\theta}{2}$.

2 If $\tan\theta = \frac{5}{12}$, find $\sin\frac{\theta}{2}$.

ISBN: 9780170425728

3 If $\sin\theta = \frac{2}{\sqrt{5}}$, find $\cos\frac{\theta}{2}$.

4 Use a double angle formula to find the exact value of $\cos\frac{\pi}{6}$.

5 Find the exact value of $\tan\frac{\pi}{8}$.

6 If $\cos\theta = -\frac{3}{5}$, find $\tan\frac{\theta}{2}$.

ISBN: 9780170425728

4 Proving identities with expressions containing double angles

- You can use similar methods to those you have used earlier.

Prove the following identities.

1 $\sec^2 \theta = \dfrac{2}{1 + \cos 2\theta}$

2 $\dfrac{1 + \cos 2\theta}{1 - \cos^2 \theta} - 2\cot^2 \theta$

3 $\dfrac{1 - \tan^2 \theta}{1 + \tan^2 \theta} = \cos 2\theta$

4 $\cos 4\theta = 8\cos^4 \theta - 8\cos^2 \theta + 1$

ISBN: 9780170425728

5 $\dfrac{\sin\theta + \sin 2\theta}{1 + \cos\theta + \cos 2\theta} = \tan\theta$

6 $2\cot 2\theta = \cot\theta - \tan\theta$

7 $\dfrac{\cos\theta + \sin\theta}{\cos\theta - \sin\theta} = \dfrac{1 + \sin 2\theta}{\cos 2\theta}$

8 $\cos 3\theta = 4\cos^3\theta - 3\cos\theta$

ISBN: 9780170425728

Sums and products

Products to sums or differences

Here are the formulae for **sines** of angle sums and differences, written the opposite way round:

$$\sin A \cos B + \cos A \sin B = \sin (A + B) \quad ①$$
$$\sin A \cos B - \cos A \sin B = \sin (A - B) \quad ②$$

Adding: ① + ②: $2\sin A \cos B = \sin (A + B) + \sin (A - B)$
Subtracting: ① – ②: $2\cos A \sin B = \sin (A + B) - \sin (A - B)$

Products — Sum and difference

Here are the formulae for **cosines** of angle sums and differences, written the opposite way round:

$$\cos A \cos B + \sin A \sin B = \cos (A - B) \quad ①$$
$$\cos A \cos B - \sin A \sin B = \cos (A + B) \quad ②$$

Adding: ① + ②: $2\cos A \cos B = \cos (A + B) + \cos (A - B)$
Subtracting: ① – ②: $2\sin A \sin B = \cos (A - B) - \cos (A + B)$

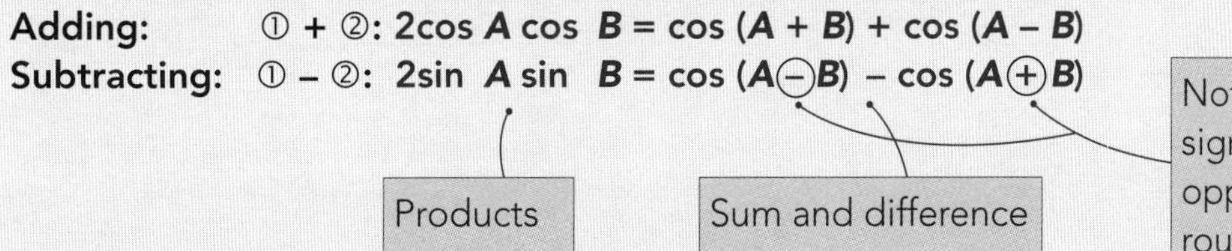

These are the new identities for converting a **product** to a **sum**:

$2\sin A \cos B = \sin (A + B) + \sin (A - B)$
$2\cos A \sin B = \sin (A + B) - \sin (A - B)$
$2\cos A \cos B = \cos (A + B) + \cos (A - B)$
$2\sin A \sin B = \cos (A - B) - \cos (A + B)$

ISBN: 9780170425728

Examples:

1 Write 2sin 58° sin 35° as a sum.

$$2\sin 58° \sin 35° = \cos(58° - 35°) - \cos(58° + 35°)$$

$$= \cos 23° - \cos 93°$$

2 Write $2\sin\frac{\pi}{6}\cos\frac{\pi}{4}$ as a sum.

$$2\sin\frac{\pi}{6}\cos\frac{\pi}{4} = \sin\left(\frac{2\pi}{12} + \frac{3\pi}{12}\right) + \sin\left(\frac{2\pi}{12} - \frac{3\pi}{12}\right)$$

$\frac{\pi}{6} < \frac{\pi}{4}$ so this produces $\sin\frac{-\pi}{12}$.

$$= \sin\frac{5\pi}{12} + \sin\frac{-\pi}{12}$$

$$= \sin\frac{5\pi}{12} - \sin\frac{\pi}{12}$$

The sine of a negative acute angle will always have a negative value.

Alternatively, when the first angle is smaller than the second angle, reverse the order of the trigonometric functions (multiplication is commutative):

$$2\sin\frac{\pi}{6}\cos\frac{\pi}{4} = 2\cos\frac{\pi}{4}\sin\frac{\pi}{6}$$

$$= \sin\left(\frac{3\pi}{12} + \frac{2\pi}{12}\right) - \sin\left(\frac{3\pi}{12} - \frac{2\pi}{12}\right)$$

$$= \sin\frac{5\pi}{12} - \sin\frac{\pi}{12}$$

3 Write $8\cos\left(\frac{\pi}{4} + 2p\right)\cos\left(\frac{\pi}{4} - 2p\right)$ as a sum.

$$8\cos\left(\frac{\pi}{4} + 2p\right)\cos\left(\frac{\pi}{4} - 2p\right) = 4\cos\frac{\pi}{2} + 4\cos 4p$$

$$= 4(0) + 4\cos 4p$$

$$= 4\cos 4p$$

 ISBN: 9780170425728

Write the following as sums or differences of single trigonometric functions.

1 $2\sin 47° \cos 26°$

2 $\cos 51° \cos 24°$

3 $2\cos \frac{\pi}{2} \sin \frac{\pi}{3}$

4 $4\sin \frac{\pi}{3} \sin \frac{\pi}{4}$

5 $\sin 66° \cos 79°$

6 $6\cos \frac{\pi}{6} \sin \frac{\pi}{4}$

ISBN: 9780170425728

7 $\frac{1}{2}\sin 3p \sin 7p$

8 $2\cos(A + B)\cos(A - B)$

9 $2\sin\left(P + \frac{\pi}{4}\right)\cos\left(P - \frac{\pi}{4}\right)$

10 $\cos\left(3\theta + \frac{\pi}{2}\right)\sin\left(3\theta - \frac{\pi}{2}\right)$

11 $2\sin(\pi + \theta)\sin(\pi - \theta)$

ISBN: 9780170425728

Sums and differences to products

Here is the formula for converting a product of a **sine and a cosine** to a sum, once again written the opposite way round:

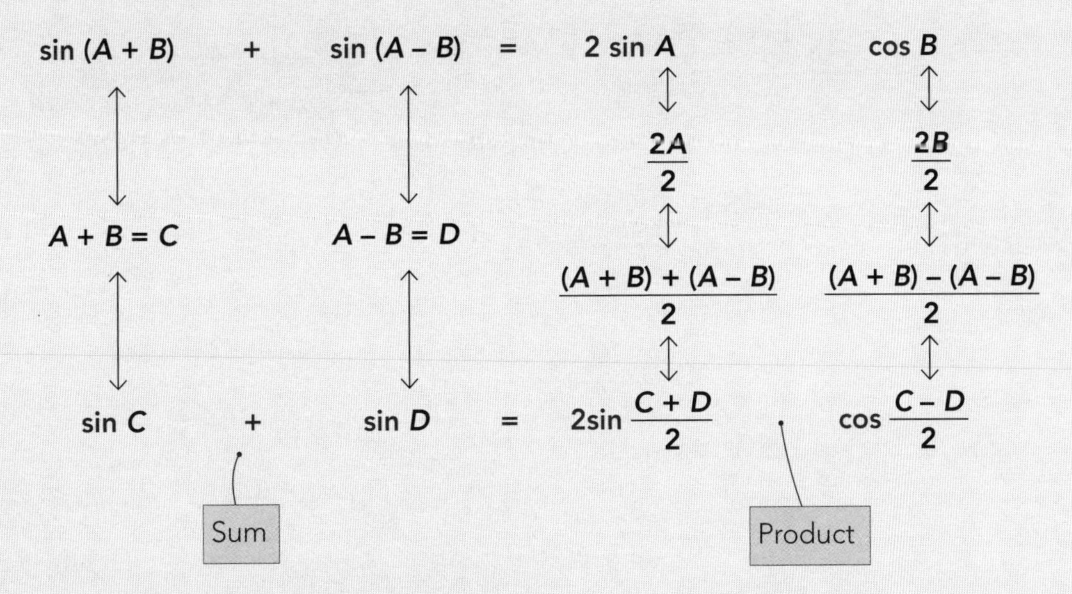

The remaining three identities below can be shown using similar reasoning.

These are the new identities for converting a **sum** to a **product**:

$$\sin C + \sin D = 2\sin\frac{C+D}{2}\cos\frac{C-D}{2}$$

$$\sin C - \sin D = 2\cos\frac{C+D}{2}\sin\frac{C-D}{2}$$

$$\cos C + \cos D = 2\cos\frac{C+D}{2}\cos\frac{C-D}{2}$$

$$\cos C - \cos D = -2\sin\frac{C+D}{2}\sin\frac{C-D}{2}$$

ISBN: 9780170425728

1 Writing sums and differences as products

Examples: Convert the following sums or differences into products.

1

$$2\cos\frac{\pi}{3} + 2\cos\frac{\pi}{2} = 2\cos\frac{\pi}{2} + 2\cos\frac{\pi}{3}$$

Reverse the order so the function of the bigger angle is first.

$$= 4\cos\frac{\frac{\pi}{2}+\frac{\pi}{3}}{2}\cos\frac{\frac{\pi}{2}-\frac{\pi}{3}}{2}$$

$$= 4\cos\frac{5\pi}{12}\cos\frac{\pi}{12}$$

2

$$\sin\left(\theta+\frac{\pi}{2}\right) + \sin\left(\theta-\frac{\pi}{2}\right) = 2\sin\frac{\left(\theta+\frac{\pi}{2}\right)+\left(\theta-\frac{\pi}{2}\right)}{2}\cos\frac{\left(\theta+\frac{\pi}{2}\right)-\left(\theta-\frac{\pi}{2}\right)}{2}$$

$$= 2\sin\frac{2\theta}{2}\cos\frac{\pi}{2}$$

$$= 2\sin\theta \times 0$$

$\cos\frac{\pi}{2} = 0$

$$= 0$$

3

$$\cos\theta + \sin\theta = \sin\left(\frac{\pi}{2}-\theta\right) + \sin\theta$$

$$= 2\sin\frac{\left(\frac{\pi}{2}-\theta\right)+\theta}{2}\cos\frac{\left(\frac{\pi}{2}-\theta\right)-\theta}{2}$$

$$= 2\sin\frac{\pi}{4}\cos\left(\frac{\pi}{4}-\theta\right)$$

Remember:

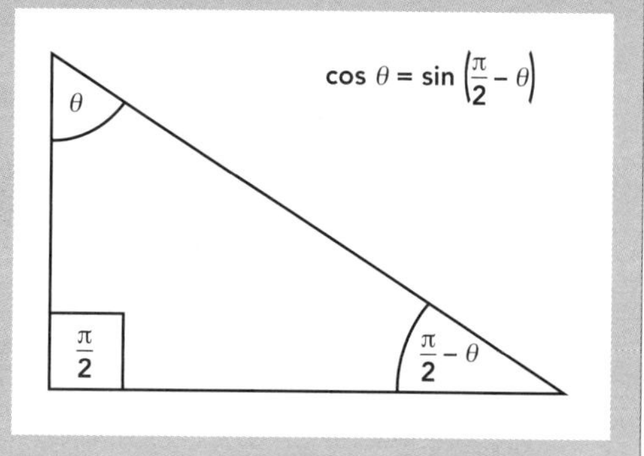

or

$$\cos\theta + \sin\theta = \cos\theta + \cos\left(\frac{\pi}{2}-\theta\right)$$

$$= 2\cos\frac{\theta+\left(\frac{\pi}{2}-\theta\right)}{2}\cos\frac{\theta-\left(\frac{\pi}{2}-\theta\right)}{2}$$

$$= 2\cos\frac{\pi}{4}\cos\left(\theta-\frac{\pi}{4}\right)$$

Similarly:

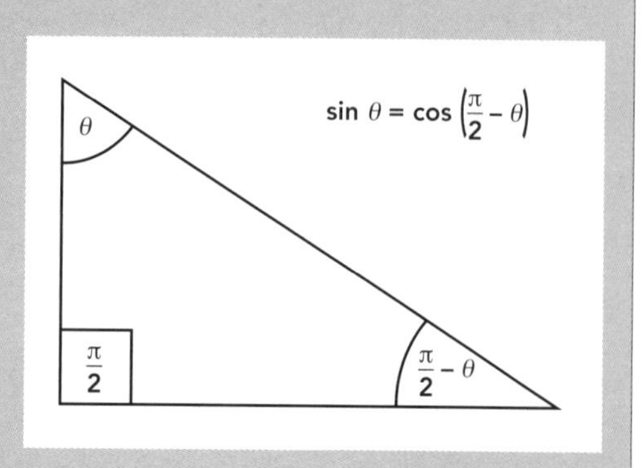

  ISBN: 9780170425728

4 $\cos\theta + 1 = \cos\theta + \cos 0$

$= 2\cos\frac{\theta + 0}{2}\cos\frac{\theta - 0}{2}$

$= 2\cos\frac{\theta}{2}\cos\frac{\theta}{2}$

$= 2\cos^2\frac{\theta}{2}$

Remember $\cos 0 = 1$

Write the following sums or differences as products.

1 $\sin A + \sin B$

2 $\cos 69° + \cos 23°$

3 $\sin\frac{\pi}{2} - \sin\frac{\pi}{4}$

4 $2\cos\frac{\pi}{4} - 2\cos\frac{\pi}{3}$

ISBN: 9780170425728

5 $\cos(\theta + 60°) + \cos(\theta - 60°)$

6 $\sin\left(2\theta + \frac{\pi}{4}\right) + \sin\left(2\theta - \frac{\pi}{4}\right)$

7 $\cos(\theta - 40°) - \cos(\theta + 40°)$

8 $\sin\theta - \cos\theta$

ISBN: 9780170425728

9 $2\cos A - 2\sin A$

10 $1 + \sin\theta$

11 $\frac{1}{2}\sin\left(2\theta + \frac{\pi}{3}\right) - \frac{1}{2}$

12 $3 + 3\cos(\theta + \pi)$

ISBN: 9780170425728

2 Proofs

Examples: Prove the following.

1 $6\sin 13y \sin 9y - 6\sin 6y \sin 2y = 6\sin 15y \sin 7y$

$$6\sin 13y \sin 9y - 6\sin 6y \sin 2y = 3((\cos 4y - \cos 22y) - (\cos 4y - \cos 8y))$$

$$= 3(\cos 4y - \cos 22y - \cos 4y + \cos 8y)$$

$$= 3\cos 8y - 3\cos 22y$$

$$= -6\sin 15y \sin(-7y)$$

$$= 6\sin 15y \sin 7y$$

2 $\cos^2 7A - \cos^2 5A = -\sin 12A \sin 2A$

$$\cos^2 7A - \cos^2 5A = \cos 7A \times \cos 7A - \cos 5A \times \cos 5A$$

$$= \frac{1}{2}(\cos 14A + \cos 0) - \frac{1}{2}(\cos 10A + \cos 0)$$

$$= \frac{1}{2}(\cos 14A - \cos 10A)$$

$$= \frac{1}{2}(-2\sin 12A \sin 2A)$$

$$= -\sin 12A \sin 2A$$

Prove the following identities.

1 $\dfrac{\sin 80° - \sin 20°}{\cos 80° + \cos 20°} = \tan 30°$

2 $\dfrac{\cos\frac{2\pi}{3} - \cos\frac{\pi}{2}}{\sin\frac{2\pi}{3} + \sin\frac{\pi}{2}} = -\tan\frac{\pi}{12}$

 ISBN: 9780170425728

3 $\sin\frac{\pi}{2} + \cos\frac{\pi}{3} = 2\sin\frac{\pi}{3}\cos\frac{\pi}{6}$

4 $\cos 3\theta + \sin\theta - \cos\theta = \sin\theta(1 - 2\sin 2\theta)$

5 $\sin 5\theta + 2\sin 3\theta + \sin\theta = 4\cos^2\theta\sin 3\theta$

6 $10\sin 4p\cos p - 10\cos 3p\sin 2p = 10\sin 2p\cos p$

7 $2\sin 13d\sin 9d - 2\sin 6d\sin 2d = 2\sin 15d\sin 7d$

8 $2\sin 5y \cos 2y - 2\sin 9y \cos 6y = -2\cos 11y \sin 4y$

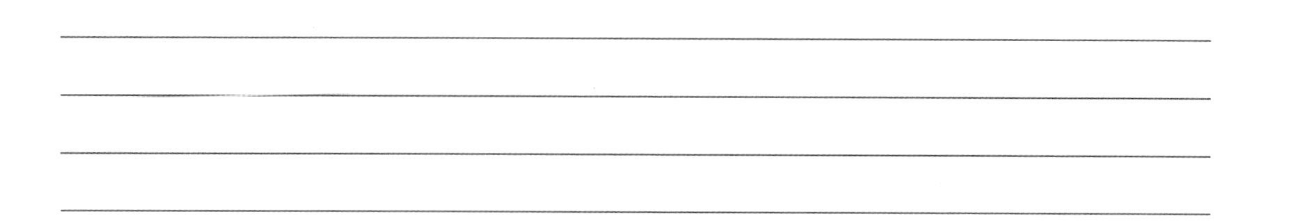

9 $\cos 7y \cos 2y - \sin 11y \sin 6y = \cos 13y \cos 4y$

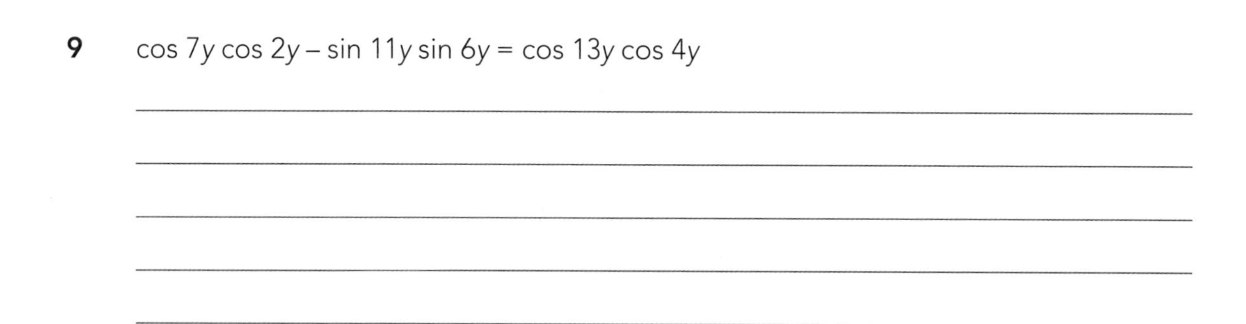

10 $2\sin^2 80° - 2\sin^2 20° = 2\sin 100° \sin 60°$

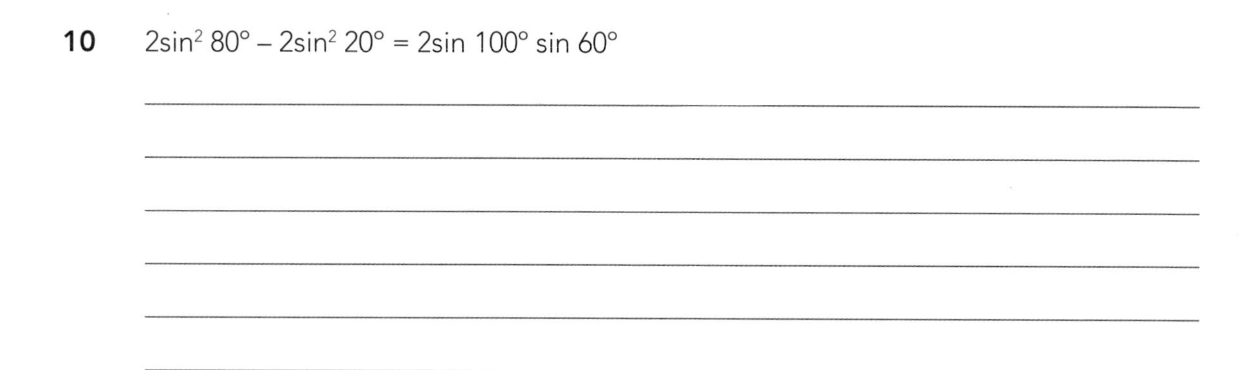

11 $\cos^2 \frac{2\pi}{3} - \sin^2 \frac{\pi}{2} = \cos \frac{7\pi}{6} \cos \frac{\pi}{6}$

ISBN: 9780170425728

Trigonometric equations

- When a trigonometric function is equal to a value, e.g. $\cos\theta = \frac{\sqrt{3}}{2}$ (or 0.866), then there can be:

  a Just one solution if, for instance, $0 \le \theta \le \frac{\pi}{2}$.

  $$\cos\theta = \frac{\sqrt{3}}{2}$$
  $$\therefore\quad \theta = \frac{\pi}{6}$$

  $\cos^{-1}$ $\cos^{-1}$

  b Many solutions if, for instance, $-2\pi \le \theta \le 4\pi$.

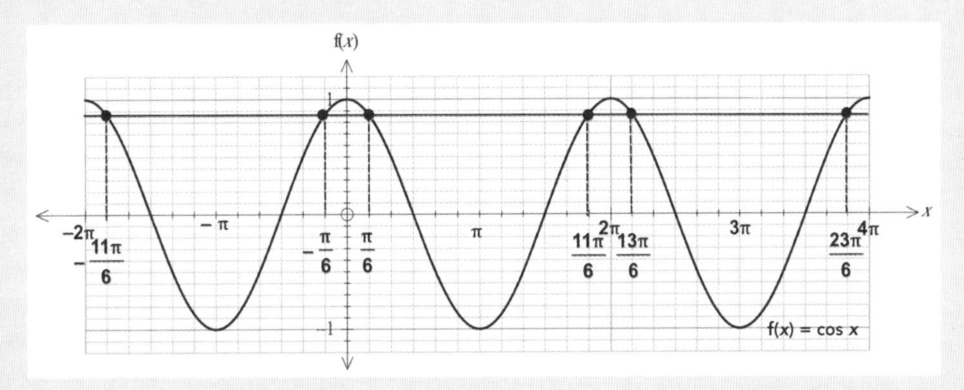

- A **general solution** to a trigonometric equation is a **formula** which gives **all** the solutions.

Formulae for general solutions

1 Cosines

Let α (alpha) be the solution that is closest to 0.

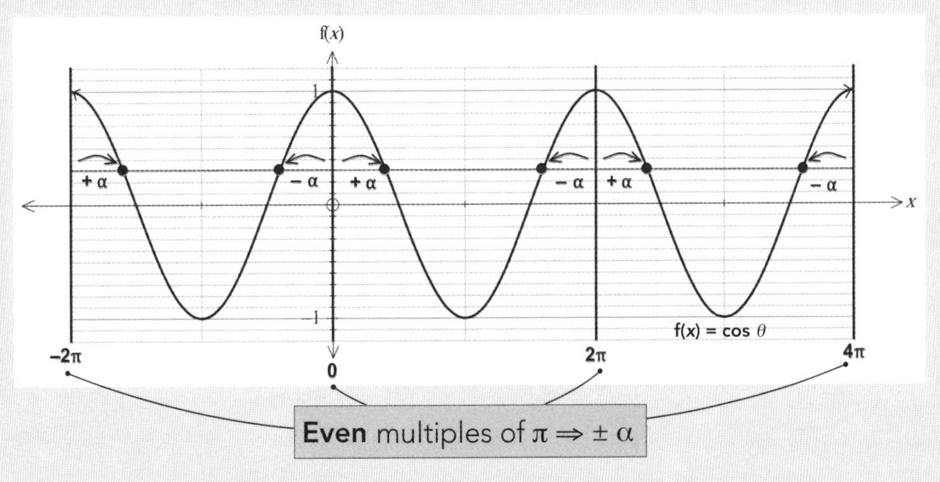

$\therefore$ $\theta = 2n\pi \pm \alpha$ **or** $\theta = 360n \pm \alpha$**, where *n* is an integer.**

2 Sines

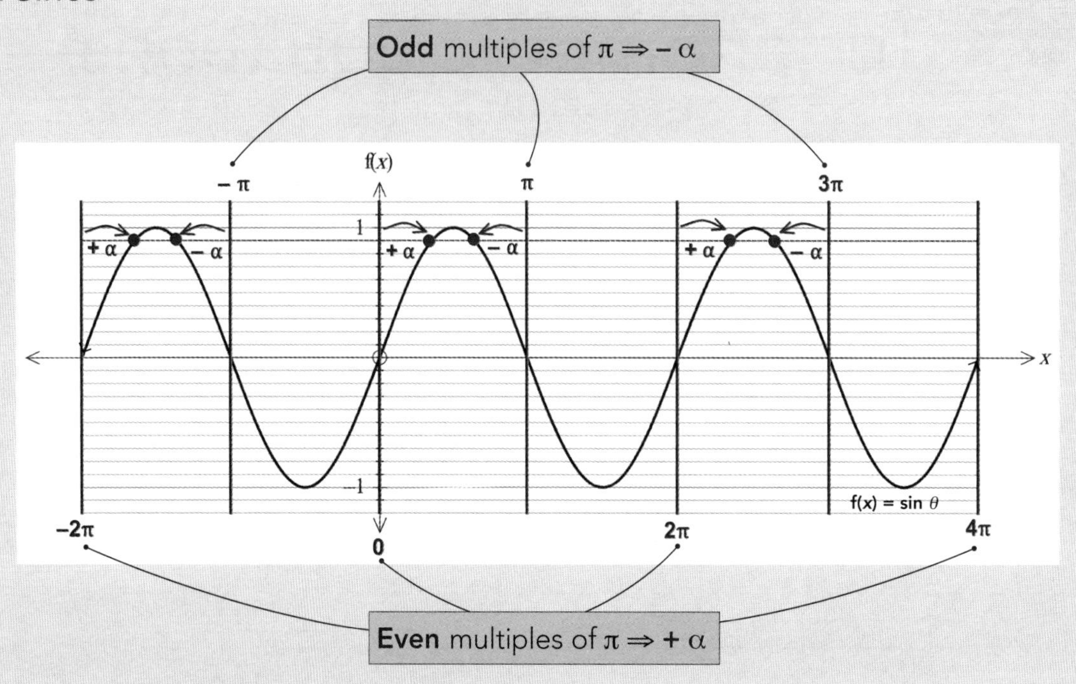

$\therefore\ \theta = n\pi + (-1)^n\,\alpha$ **or** $\theta = 180n + (-1)^n\,\alpha$**, where *n* is an integer.**

3 Tangents

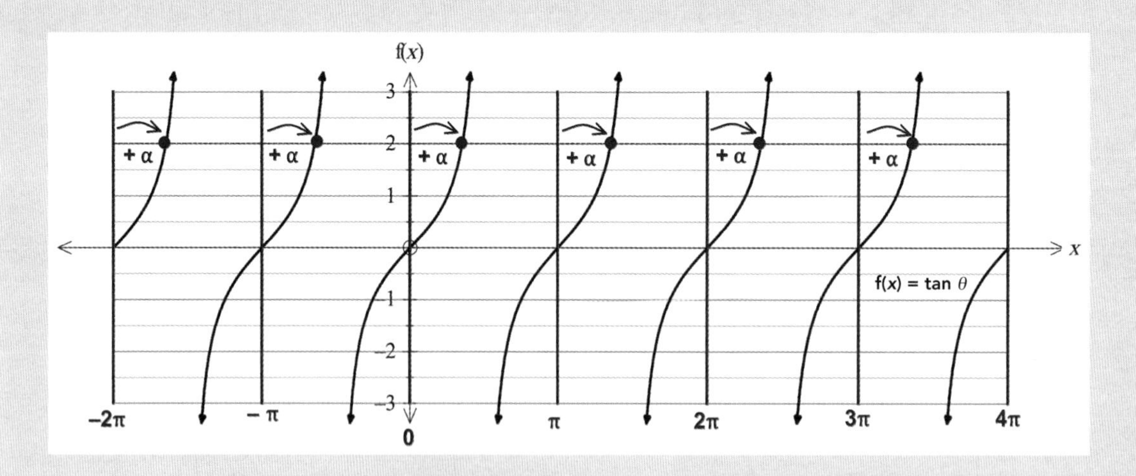

All multiples of π ⇒ + α

$\therefore\ \theta = n\pi + \alpha$ **or** $\theta = 180n + \alpha$**, where *n* is an integer.**

ISBN: 9780170425728

Basic equations

Examples:

1 Let $\sin\theta = 0.3420$ and $-360° \le \theta \le 360°$.

Range in degrees ⇒ answer in degrees.

a Find the general solution.

Step 1: Find α.

sin^{-1} $\sin\theta = 0.3420$ sin^{-1}

$\theta = 20°$

$\therefore$ Let $\alpha = 20°$

Step 2: Apply the appropriate general solution formula.

$$\theta = 180n + (-1)^n\alpha$$

$$\therefore \quad \boldsymbol{\theta = 180n + (-1)^n \times 20°}$$

b Find all the solutions to $\sin\theta = 0.3420$ for $-360° < \theta < 360°$.

$n = -2 \Rightarrow \quad \theta = -2 \times 180 + (-1)^{-2} \times 20° = -360° + 20°$
$= -340°$

$n = -1 \Rightarrow \quad \theta = -1 \times 180 + (-1)^{-1} \times 20° = -180° - 20°$
$= -200°$

$n = 0 \Rightarrow \quad \theta = 0 \times 180 + (-1)^{0} \times 20° = 0° + 20°$
$= 20°$

$n = 1 \Rightarrow \quad \theta = 1 \times 180 + (-1)^{1} \times 20° = 180° - 20°$
$= 160°$

$n = 2 \Rightarrow \quad \theta = 2 \times 180 + (-1)^{2} \times 20° = 360° + 20°$
$= 380°$, which is outside the range.

$\therefore$ **Solutions are: –340°, –200°, 20° and 160°.**

2 Let $\tan\theta = -\sqrt{3}$ and $-2\pi \le \theta \le 3\pi$.

Range in radians ⇒ answer in radians.

a Find the general solution.

Step 1:

$$\tan\theta = -\sqrt{3}$$

$$\theta = -\frac{\pi}{3}$$

$$\therefore \text{Let } \alpha = -\frac{\pi}{3}$$

Step 2:

$$\theta = n\pi + \alpha$$

$$\therefore \boldsymbol{\theta = n\pi - \frac{\pi}{3}}$$

ISBN: 9780170425728

b Find all the exact solutions to $\tan\theta = \sqrt{3}$ for $-2\pi < \theta < 2\pi$.

$n = -1 \Rightarrow \quad \theta = -\pi - \frac{\pi}{3} = -\frac{4\pi}{3}$

$n = 0 \Rightarrow \quad \theta = 0\pi - \frac{\pi}{3} = -\frac{\pi}{3}$

$n = 1 \Rightarrow \quad \theta = \pi - \frac{\pi}{3} = \frac{2\pi}{3}$

$n = 2 \Rightarrow \quad \theta = 2\pi - \frac{\pi}{3} = \frac{5\pi}{3}$

$\therefore$ **Solutions are:** $-\frac{4\pi}{3}, -\frac{\pi}{3}, \frac{2\pi}{3}$ **and** $\frac{5\pi}{3}$.

Notice that there is a pattern in the answers. The next smaller answer would be $-\frac{7\pi}{3}$, which would be outside the range.

The next bigger answer would be $\frac{8\pi}{3}$, which would also be outside the range.

3 Let $2\cos\theta + 3 = 4.4142$ and $0 \leq \theta \leq 3\pi$.

a Find the general solution.

Step 1: $2\cos\theta = 4.4142 - 3$

$\cos\theta = 0.7071$

$\theta = 45° = \frac{\pi}{4}$

$\therefore$ Let $\alpha = \frac{\pi}{4}$

Rearrangement needed in order to find α.

Step 2: $\theta = 2n\pi \pm \alpha$

$\boldsymbol{\theta = 2n\pi \pm \frac{\pi}{4}}$

b Find all the exact solutions to $2\cos\theta + 3 = 4.4142$ for $0 < \theta < 3\pi$.

$n = 0 \Rightarrow \quad \theta = 0\pi - \frac{\pi}{4} = -\frac{\pi}{4}$ Outside the range.

$\theta = 0\pi + \frac{\pi}{4} = \frac{\pi}{4}$

$n = 1 \Rightarrow \quad \theta = 2\pi - \frac{\pi}{4} = \frac{7\pi}{4}$

$\theta = 2\pi + \frac{\pi}{4} = \frac{9\pi}{4}$

From the pattern, the next solution would be $\frac{15\pi}{4}$, which would be outside the range.

$\therefore$ **Solutions are:** $\frac{\pi}{4}, \frac{7\pi}{4}$ **and** $\frac{9\pi}{4}$.

 ISBN: 9780170425728

4 Let $\sin \theta = 2\cos \theta$ where $0 \leq \theta \leq 2\pi$.

a Find the general solution.

Step 1: $\dfrac{\sin \theta}{\cos \theta} = 2$

$\tan \theta = 2$

$\therefore$ Let $\alpha = 1.1071$

Step 2: $\theta = n\pi + \alpha$

$\boldsymbol{\theta = n\pi + 1.1071}$

b Find all the solutions to $\sin \theta = 2\cos \theta$ where $0 \leq \theta \leq 2\pi$.

$n = 0 \Rightarrow \theta = 1.1071$
$n = 1 \Rightarrow \theta = 4.2487$

$\therefore$ **Solutions are 1.1071 and 4.2487.**

Find general and particular solutions for the following. For questions marked with a *, your particular solutions should be exact.

1 $\cos \theta = 0.4226$ when $-360° \leq \theta \leq 400°$

2 $\sin \theta = 0.9848$ when $0° \leq \theta \leq 540°$

ISBN: 9780170425728

3 $\tan \theta = 0.5317$ when $-180° \leq \theta \leq 300°$

4 $\sin \theta = 0.548$ when $-\pi \leq \theta \leq 2\pi$
(Hint: Change your calculator to radians.)

5 $\cos \theta = 0.7518$ when $0 \leq \theta \leq 4\pi$

6 $\tan \theta = 0.347$ when $-3\pi \leq \theta \leq \pi$

***7** $\cos \theta = \frac{1}{\sqrt{2}}$ when $-\pi \leq \theta \leq 2\pi$

***8** $\tan \theta = -\sqrt{3}$ when $0 \leq \theta \leq 4\pi$

ISBN: 9780170425728

⋆ **9** $\sin\theta = -\frac{1}{2}$ when $-2\pi \le \theta \le 2\pi$

10 $4\cos\theta = 3.8756$ when $-\pi \le \theta \le 3\pi$

11 $\tan\theta + 2 = 2.5206$ when $\pi \le \theta \le 3\pi$

12 $3\sin\theta - 2 = 0.673$ when $100° \le \theta \le 400°$

⋆ **13** $\sin\theta = \frac{\cos\theta}{\sqrt{3}}$ when $0 \le \theta \le 3\pi$

⋆ **14** $\cos\theta = -\sin\theta$ when $0° \le \theta \le 720°$

ISBN: 9780170425728

Rearrangements of general solutions

When you are given trigonometric functions of another function, rearrange in order to find the general solution.

Examples:

1 Solve $\cos 4\theta = 0.3090$ when $0° \le \theta \le 200°$.

$\cos 4\theta = 0.3090 \Rightarrow \alpha = 72°$

$\therefore\ 4\theta = 360n \pm 72°$ — **Rearrange**

$\boldsymbol{\theta = 90n \pm 18°}$

$n = 0 \Rightarrow$ $\theta = 90 \times 0 - 18° = -18°$ Out of range

$\theta = 90 \times 0 + 18° = 18°$

$n = 1 \Rightarrow$ $\theta = 90 \times 1 - 18° = 72°$

$\theta = 90 \times 1 + 18° = 108°$

$n = 2 \Rightarrow$ $\theta = 90 \times 2 - 18° = 162°$

$\theta = 90 \times 2 + 18° = 198°$

$\therefore$ **Solutions are: 18°, 72°, 108°, 162° and 198°.**

2 Solve $\sin(2\theta + 1) = 0.3986$ for the range $0 \le \theta \le 6$.

$\sin(2\theta + 1) = 0.3986 \Rightarrow \alpha = 0.41$

$\therefore\ 2\theta + 1 = n\pi + (-1)^n \times 0.41$ — **Rearrange**

$2\theta = n\pi + (-1)^n \times 0.41 - 1$ — **Rearrange**

$\boldsymbol{\theta = \frac{n\pi}{2} + (-1)^n \times 0.205 - 0.5}$

$n = 1 \Rightarrow$ $\theta = \frac{\pi}{2} - 0.205 - 0.5 = 0.8658$

$n = 2 \Rightarrow$ $\theta = \pi + 0.205 - 0.5 = 2.8466$

$n = 3 \Rightarrow$ $\theta = \frac{3\pi}{2} - 0.205 - 0.5 = 4.0074$

$n = 4 \Rightarrow$ $\theta = 2\pi + 0.205 - 0.5 = 5.9882$

$\therefore$ **Solutions are: 0.8658, 2.8466, 4.0074 and 5.9882.**

ISBN: 9780170425728

Find general and particular solutions for each of the following.

1 $\tan 4\theta = 1.28$ for $-90° \leq \theta \leq 90°$

2 $\cos 2\theta = 0.0349$ for $0° \leq \theta \leq 120°$

3 $\sin 8\theta = 0.8290$ for $-30° \leq \theta \leq 30°$

4 $\tan 4\theta = 0.2027$ for $-\frac{\pi}{2} \leq \theta \leq \frac{\pi}{2}$

5 $\cos 3\theta = 0.9359$ for $0 \leq \theta \leq \pi$

6 $\tan(2\theta + 1) = 3.6021$ for $-\pi \leq \theta \leq \pi$

7 $\cos(2\theta + 1) = 0.8419$ for $0 \leq \theta \leq \pi$

8 $\sin(3\theta + 2) = 0.5155$ for $0 \leq \theta \leq \pi$

ISBN: 9780170425728

Solving trigonometric equations by factorising

Remember that $a \times b = 0 \Rightarrow$ either $a = 0$ or $b = 0$ (or both).

a By factorising directly or using the Pythagorean identities

Examples: Solve the following.

1 $\sin^2\theta + 2\cos\theta + 2 = 0$ when $0 \le \theta \le 4\pi$

$(1 - \cos^2\theta) + 2\cos\theta + 2 = 0$

Replace $\sin^2\theta$ by $1 - \cos^2\theta$

$\cos^2\theta - 2\cos\theta - 3 = 0$

$(\cos\theta - 3)(\cos\theta + 1) = 0$

Either $\cos\theta = 3$, which is not possible,

or $\cos\theta = -1$

$\alpha = \pi$

$\boldsymbol{\theta = 2n\pi \pm \pi}$

$n = 0 \Rightarrow \theta = \pi$

$n = 1 \Rightarrow \theta = 3\pi$

$\therefore$ **Solutions are π and 3π.**

2 $2\sin\theta\cos\theta = \sin\theta$ where $0 \le \theta \le 2\pi$

$2\sin\theta\cos\theta - \sin\theta = 0$

Do **not** divide both sides by a common factor (in this case $\sin\theta$) so that one function is eliminated. If you do, some solutions may not be found.
Dividing by $\sin\theta$ in this equation would result in $\cos\theta = \frac{1}{2}$ only, and the solutions 0, π, and 2π would not be found.

$\sin\theta(2\cos\theta - 1) = 0$

Either $\sin\theta = 0$, so $\theta = 0, \pi$ or 2π

or $\cos\theta = \frac{1}{2}$ so $\theta = \frac{\pi}{3}$ or $\frac{5\pi}{3}$.

$\therefore$ **Solutions are: $0, \frac{\pi}{3}, \pi, \frac{5\pi}{3}$ and 2π.**

3 $2\sin\theta + \operatorname{cosec}\theta - 3 = 0$ where $0 \le \theta \le 2\pi$

$2\sin\theta + \frac{1}{\sin\theta} - 3 = 0$

Multiply everything by $\sin\theta$.

$2\sin^2\theta + 1 - 3\sin\theta = 0$

$2\sin^2\theta - 3\sin\theta + 1 = 0$

$(2\sin\theta - 1)(\sin\theta - 1) = 0$

Either $\sin\theta = \frac{1}{2} \Rightarrow \theta = \frac{\pi}{6}$ or $\frac{5\pi}{6}$ or $\sin\theta = 1 \Rightarrow \theta = \frac{\pi}{2}$.

$\therefore$ **Solutions are: $\frac{\pi}{6}, \frac{\pi}{2}$ and $\frac{5\pi}{6}$.**

ISBN: 9780170425728

4 $3\sin^2\theta - \sin\theta\cos\theta = 1$ for $0 \le \theta \le 2\pi$

$3\sin^2\theta - \sin\theta\cos\theta - 1 = 0$

$3\sin^2\theta - \sin\theta\cos\theta - (\sin^2\theta + \cos^2\theta) = 0$ — Replace 1 by $\sin^2\theta + \cos^2\theta$

$3\sin^2\theta - \sin\theta\cos\theta - \sin^2\theta - \cos^2\theta = 0$

$2\sin^2\theta - \sin\theta\cos\theta - \cos^2\theta = 0$

$(2\sin\theta + \cos\theta)(\sin\theta - \cos\theta) = 0$

Either

$2\sin\theta = -\cos\theta$

$2\tan\theta = -1$

$\tan\theta = -\frac{1}{2}$

$\alpha = -0.4636$

$\boldsymbol{\theta = n\pi - 0.4636}$

$n = 1 \Rightarrow \theta = 2.6780$

$n = 2 \Rightarrow \theta = 5.8196$

or

$\sin\theta = \cos\theta$

$\tan\theta = 1$

$\alpha = \frac{\pi}{4}$

$\boldsymbol{\theta = n\pi + \frac{\pi}{4}}$

$n = 0 \Rightarrow \theta = \frac{\pi}{4}$

$n = 1 \Rightarrow \theta = \frac{5\pi}{4}$

$\therefore$ **Solutions are: $\frac{\pi}{4}$, 2.6780, $\frac{5\pi}{4}$ and 5.8196.**

Find solutions for each of the following.

1 $\cos^2\theta - \cos\theta = 0$ where $0 \le \theta \le 2\pi$

2 $\tan^2\theta + 5\tan\theta = 0$ where $0° \le \theta \le 360°$

ISBN: 9780170425728

3 $4\cos^2\theta - 3\cos\theta - 1 = 0$ where $0 \le \theta \le 2\pi$

4 $\sin^2\theta - \cos^2\theta = 0$ where $0 \le \theta \le 2\pi$

5 $\tan^2\theta = 3\tan\theta$ where $0 \le \theta \le 2\pi$

6 $\tan\theta\cos^2\theta = -3\tan\theta$ where $0 \le \theta \le 4\pi$

7 $\cos^2\theta + 3\cos\theta = 0$ where $0 \le \theta \le 4\pi$

8 $\cos\theta + 2\sec\theta - 3 = 0$ where $0 \le \theta \le 4\pi$

ISBN: 9780170425728

9 $\tan\theta - 6\cot\theta - 5 = 0$ where $0 \le \theta \le 2\pi$

10 $2\cos^2\theta - 4\sin\theta\cos\theta = -1$ where $0 \le \theta \le 2\pi$

b Using the double angle and other identities

Use double angle identities to convert functions such as $\sin 2\theta$, $\cos 2\theta$, etc. into functions of single angles.

Find solutions for each of the following.

1 $\cos 2\theta + \cos\theta = 0$ where $0 \le \theta \le 2\pi$

2 $2\cos^2\theta - \sin 2\theta = 0$ where $0 \le \theta \le 2\pi$

 ISBN: 9780170425728

3 $\cos\theta\sin\theta - \cos 2\theta = -1$ where $0 \leq \theta \leq 2\pi$

4 $\cos^2\theta - \sin^2 2\theta = 0$ where $0 \leq \theta \leq 2\pi$

5 $\tan 2\theta = 4\tan\theta$ where $0 \leq \theta \leq \pi$

6 $5\cos^2\theta + \sin 2\theta = 1 - \sin^2\theta$ where $0 \leq \theta \leq 2\pi$

ISBN: 9780170425728

c By converting sums and differences to products

Use the sums identities to convert sums to products.

Find solutions for each of the following.

1 $\sin 3\theta + \sin \theta = 0$ where $0 \le \theta \le 2\pi$

2 $\cos 8\theta - \cos 2\theta = 0$ where $0 \le \theta \le \pi$

3 $\cos \theta - \sin \theta = 0$ where $0 \le \theta \le 2\pi$

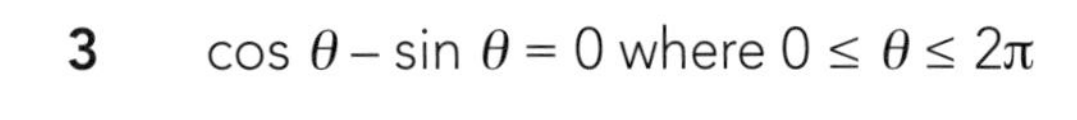

4 $\dfrac{\sin 3\theta + \sin \theta}{\cos \theta} = 1$ where $0 \le \theta \le \pi$

5 $\sin 3\theta - \sin \theta = 3\cos 2\theta$ where $0 \le \theta \le 2\pi$

6 $\cos 4\theta - \cos 2\theta = \sin 3\theta$ where $0 \le \theta \le \pi$

 ISBN: 9780170425728

d Solving equations by squaring

- Squaring identities can result in extra **invalid solutions**.
- You need to **test** each solution in the original equations to check if it is valid.

Example:

Solve $2\sin\theta = 1 - \cos\theta$ where $0 \le \theta \le 3\pi$

$$4\sin^2\theta = 1 - 2\cos\theta + \cos^2\theta$$ Square both sides.

$$4(1 - \cos^2\theta) = 1 - 2\cos\theta + \cos^2\theta$$

$$4 - 4\cos^2\theta = 1 - 2\cos\theta + \cos^2\theta$$

$$5\cos^2\theta - 2\cos\theta - 3 = 0$$

$$(5\cos\theta + 3)(\cos\theta - 1) = 0$$

$\therefore$ Either $\cos\theta = -0.6$

$\alpha = 2.2143$

$\boldsymbol{\theta = 2n\pi \pm 2.2143}$

$n = 0 \Rightarrow \theta = 2.2143$

$n = 1 \Rightarrow \theta = 2\pi - 2.2143 = 4.0689$

$\theta = 2\pi + 2.2143 = 8.4975$

or $\cos\theta = 1$

$\alpha = 0$

$\boldsymbol{\theta = 2n\pi \pm 0}$

$\theta = 0, 2\pi$

$\therefore$ Solutions **appear** to be: 0, 2.2143, 4.0689, 2p and 8.4975.

Test each solution:

$\theta = 0$	$2\sin 0 = 1 - \cos 0$	✓
$\theta = 2.2143$	$2\sin 2.2143 = 1 - \cos 2.2143$	✓
$\theta = 4.0689$	$2\sin 4.0689 \neq 1 - \cos 4.0689$	✗
$\theta = 2\pi$	$2\sin 2\pi = 1 - \cos 2\pi$	✓
$\theta = 8.4975$	$2\sin 8.4975 = 1 - \cos 8.4975$	✓

$\therefore$ **Solutions are: 0, 2.2143, 2π and 8.4975.**

ISBN: 9780170425728

Find solutions for each of the following.

1 $3\cos\theta = 2 - \sin\theta$ where $0 \leq \theta \leq 3\pi$

2 $1 + 3\sin\theta = 2\cos\theta$ where $0 \leq \theta \leq 3\pi$

 ISBN: 9780170425728

Using a graphics calculator to solve trigonometric equations (very useful for checking answers)

Example: Solve 2sin θ = 1 – cos θ where 0 ≤ θ ≤ 3π.

Rearrange so that 0 is on one side: 2sin θ – 1 + cos θ = 0

Check that your calculator is set to radians: see page 44.

Go to Graph on the main menu: ⇒ EXE

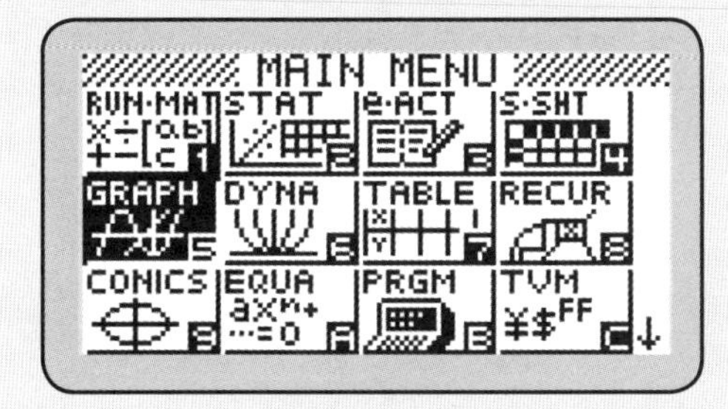

Enter the left side of the equation into Y1: ⇒ 2sin x – 1 + cos x

⇒ EXE

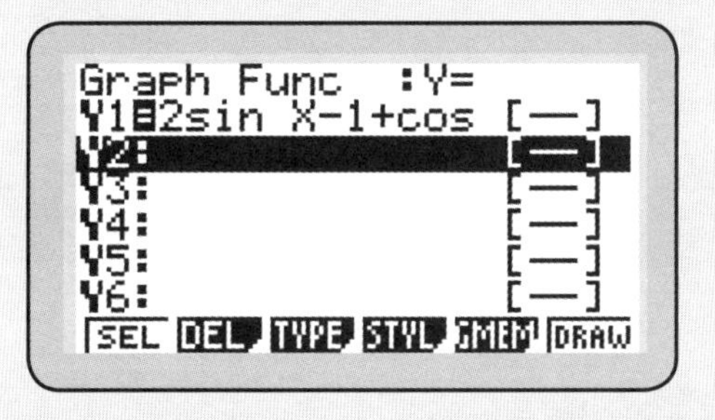

Set the view window: ⇒ Shift F3 (V-window)

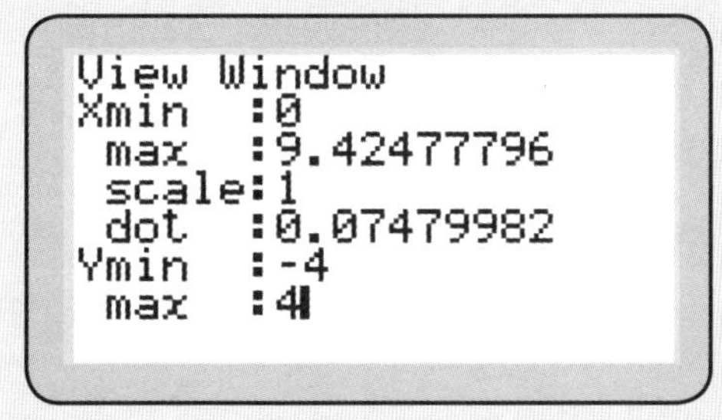

For this graph:

1. Set Xmin to 0. EXE
2. Set Xmax to 3π. EXE
3. ∇ ∇
4. Change the Ymin to -4. EXE
5. Change the Ymax to 4. EXE EXE

ISBN: 9780170425728

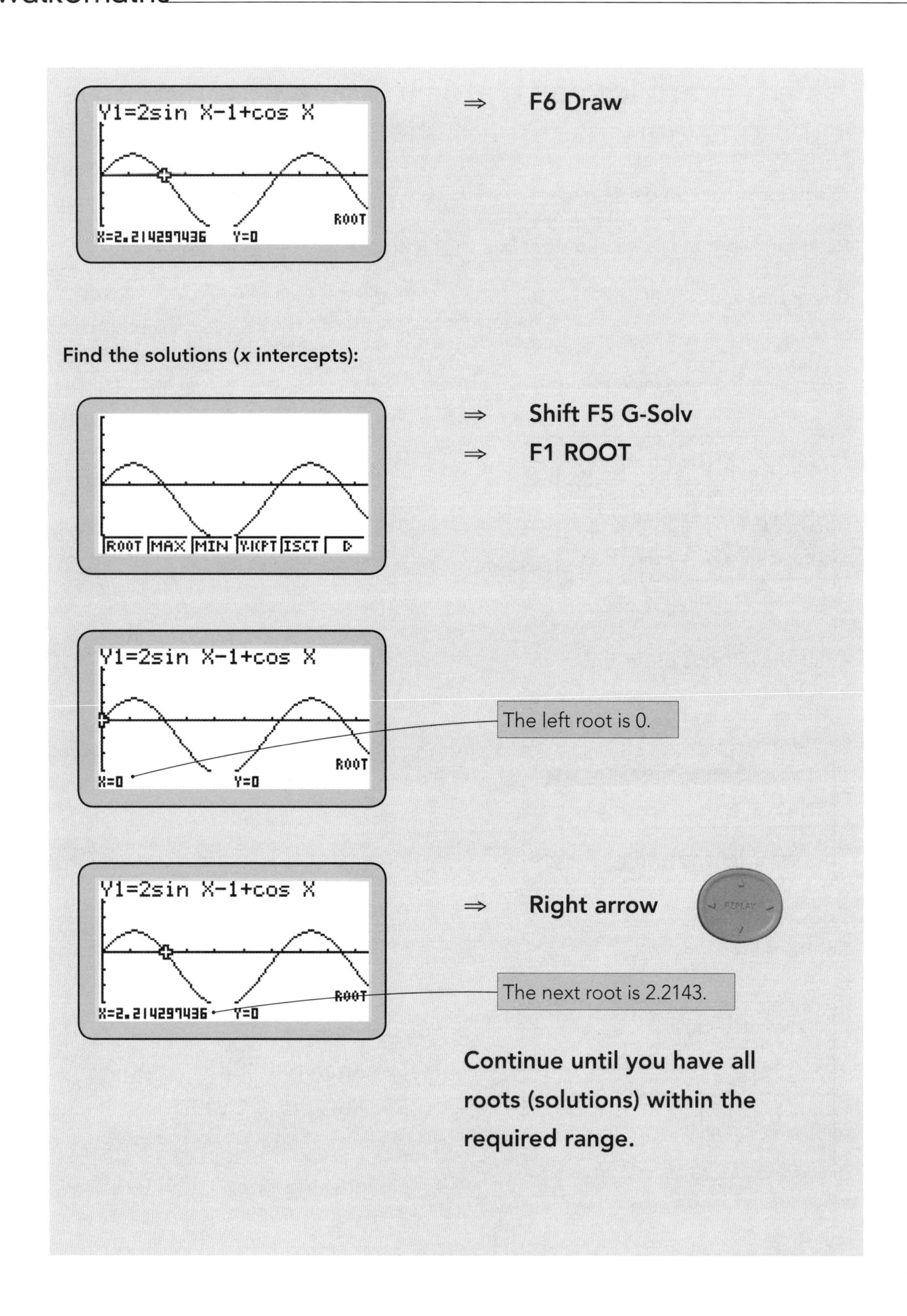

ISBN: 9780170425728

You can also find other features of the graph.

The maximum:

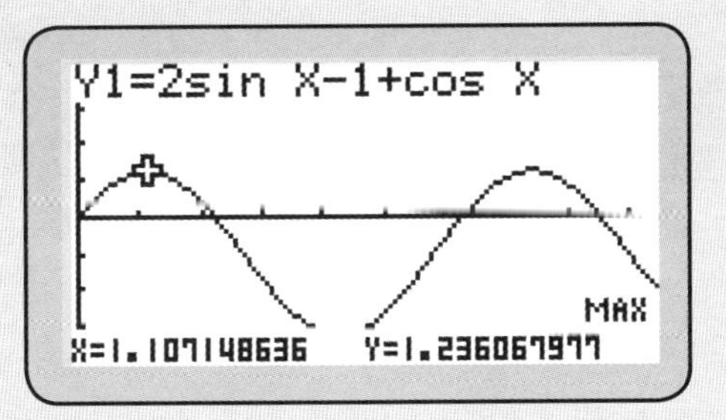

⇒ **Shift F5 G-Solv**

⇒ **F2 MAX**

The minimum:

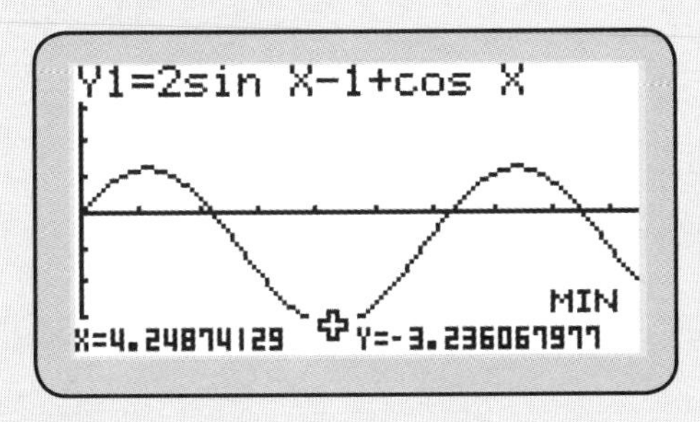

⇒ **Shift F5 G-Solv**

⇒ **F3 MIN**

***y* intercept:**

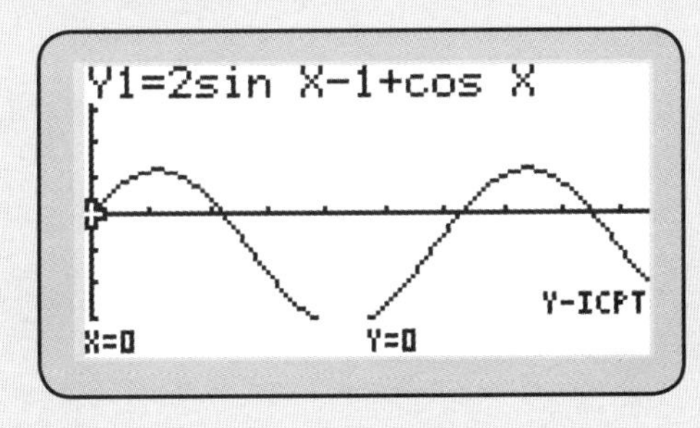

⇒ **Shift F5 G-Solv**

⇒ **F4 Y-ICPT**

Points of intersection – where two equations have been entered:

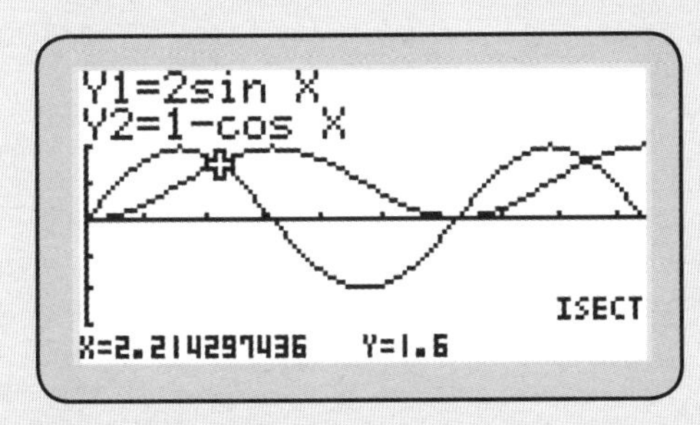

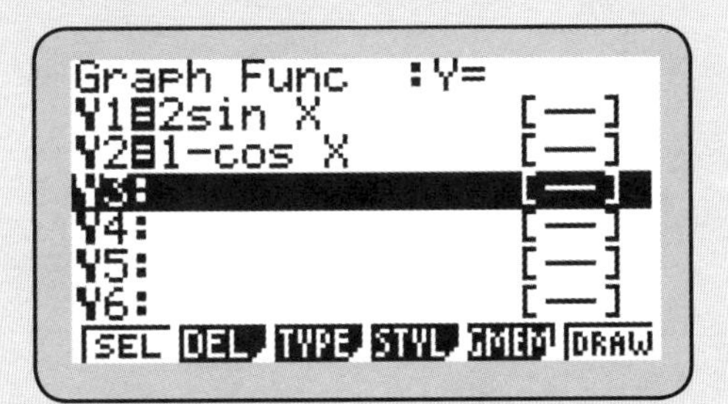

⇒ **Shift F5 G-Solv**

⇒ **F5 ISCT**

Note: When entering an equation such as

$$y = 5\sin \frac{\pi}{2}(x + 2) - 4$$

1. The entire expression which you want to find the trigonometric function of must be bracketed.
2. The calculator does not handle fractions of π.

∴ It must be written as

$y = 5\sin\ \mathbf{(1.5708}(x + 2)\mathbf{)} - 4$

or $y = 5\sin\ \mathbf{(0.5\pi}(x + 2)\mathbf{)} - 4$

ISBN: 9780170425728

Putting it together

You will be expected to be able to use combinations of anything covered so far.

Example:

Tides

In order to get out to the open sea, Ben has to sail his yacht over a sand bar, so he needs an equation to calculate the depth of water at any particular time.

The time between high tides is 12.5 hours, and the first high tide of the weekend will occur at 4 am on Saturday. The maximum water depth over the bar is 3.2 m and the minimum depth is 0.8 m.

a Use the model $y = a\cos b(x \pm c) \pm d$ to write an equation for the depth of water over the bar, where y represents depth (m) and x represents time (h) after midnight on Friday.

a = amplitude = $\frac{3.2 - 0.8}{2} = 1.2$ **b** = frequency = $\frac{2\pi}{12.5}$

c = horizontal shift = -4 **d** = vertical shift = $\frac{3.2 + 0.8}{2} = 2$

$\therefore$ **Equation is** $\mathbf{y = 1.2\cos \frac{2\pi}{12.5}(x - 4) + 2}$

b At what times will low tide occur on Saturday and Sunday?

Low tide $\Rightarrow y = 1.2\cos \frac{2\pi}{12.5}(x - 4) + 2$, so $1.2\cos \frac{2\pi}{12.5}(x - 4) + 2 = 0.8$

$$1.2\cos \frac{2\pi}{12.5}(x - 4) + 2 = 0.8$$

$$1.2\cos \frac{2\pi}{12.5}(x - 4) = -1.2$$

$$\cos \frac{2\pi}{12.5}(x - 4) = -1$$

$$\alpha = \pi$$

$$\frac{2\pi}{12.5}(x - 4) = 2n\pi \pm \pi$$

$$(x - 4) = 12.5n \pm 6.25$$

$$\mathbf{x = 12.5n \pm 6.25 + 4}$$

$n = 0 \Rightarrow$ $x = 10.25$, which is **10:15 am on Saturday**.

$n = 1 \Rightarrow$ $x = 12.5 - 6.25 + 4 = 10.25$, which is the same.

$x = 12.5 + 6.25 + 4 = 22.75$, which is **10:45 pm on Saturday**.

$n = 2 \Rightarrow$ $x = 25 - 6.25 + 4 = 22.75$, which is the same.

$x = 25 + 6.25 + 4 = 35.25$, which is **11:15 am on Sunday**.

$n = 3 \Rightarrow$ $x = 37.5 - 6.25 + 4 = 35.25$, which is the same.

$x = 37.5 + 6.25 + 4 = 47.75$, which is **11:45 pm on Sunday**.

 ISBN: 9780170425728

c Ben needs 2 m of water over the sand bar if he is to cross it safely. Find a formula that would enable him to calculate when this depth occurs.

$$1.2\cos\frac{2\pi}{12.5}(x-4)+2=2$$

$$1.2\cos\frac{2\pi}{12.5}(x-4)=0$$

$$\cos\frac{2\pi}{12.5}(x-4)=0$$

$$\alpha=\frac{\pi}{2}$$

$$\frac{2\pi}{12.5}(x-4)=2n\pi\pm\frac{\pi}{2}$$

$$x-4=12.5n\pm3.125$$

$$\boldsymbol{x=12.5n\pm3.125+4}$$

d Between what times on Saturday will there be at least 2 m over the bar?

There is exactly 2 m of water over the sand bar at:

$n=0\Rightarrow$ $x=0.875$, which is 0:53 am on Saturday.
$x=7.125$, which is 7:08 am on Saturday.
$n=1\Rightarrow$ $x=13.375$, which is 1:23 pm on Saturday.
$x=19.625$, which is 7:38 pm on Saturday.

Sketch the graph:

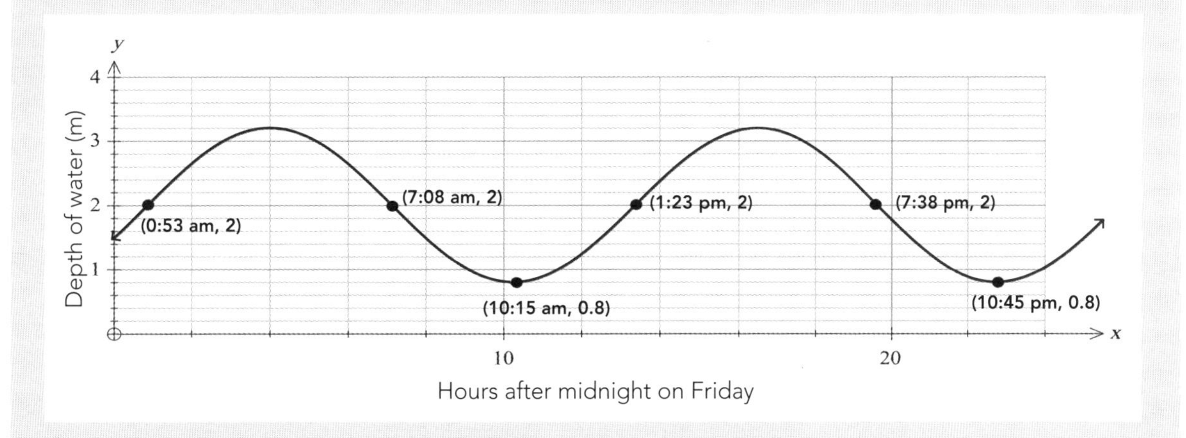

∴ **On Saturday there should be at least 2 m of water over the sand bar between 0:53 am and 7:08 am and between 1:23 pm and 7:38 pm.**

Answer the following, showing your reasoning.

1 Hannah has a crystal hung in her bedroom window. When she pulls it down, it bounces up and down. Let h represent its height in centimetres above the surface of her window sill of the lowest point on the crystal, and t represent time in seconds. The graph shows its motion.

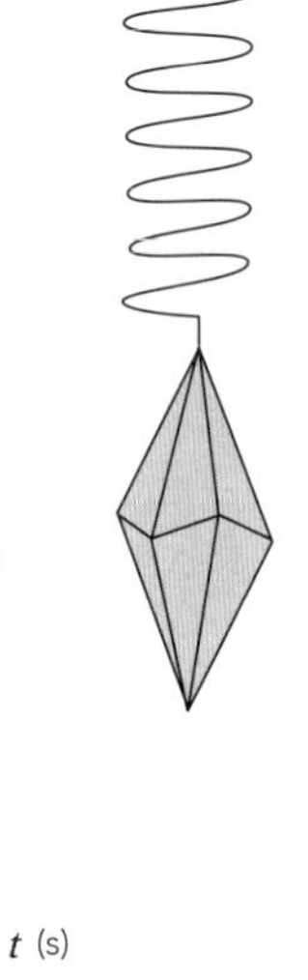

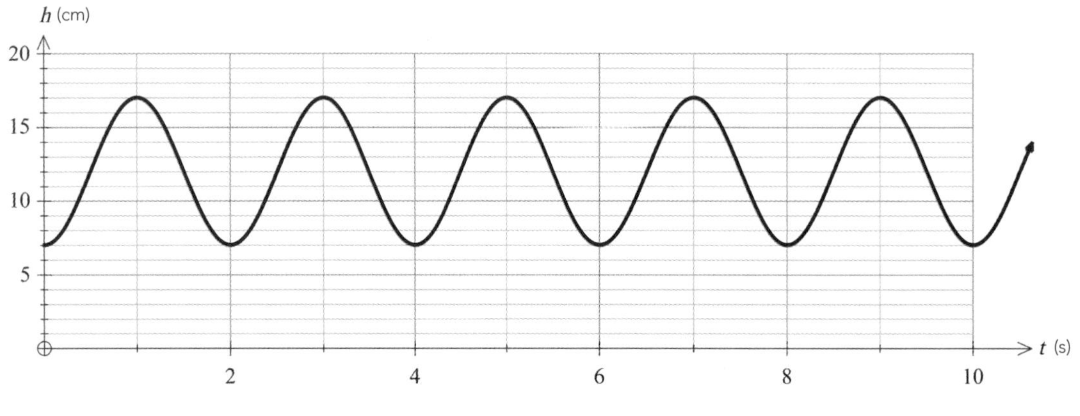

a Use the information on the graph to write an equation that models the motion of the crystal. Use $y = a\cos b(x \pm c) \pm d$ as a basis for your model and show your reasoning.

b Find a general solution for when the lowest point on the crystal is exactly 9 cm above the surface of the window sill.

c Calculate when, during the first two seconds, the lowest point on the crystal is exactly 9 cm above the surface of the window sill.

ISBN: 9780170425728

2 a Draw the graph of $y = 5\cos(x - 0.6435)$ between 0 and 2π. Include the coordinates of all significant points, along with any reasoning and working required to find them.

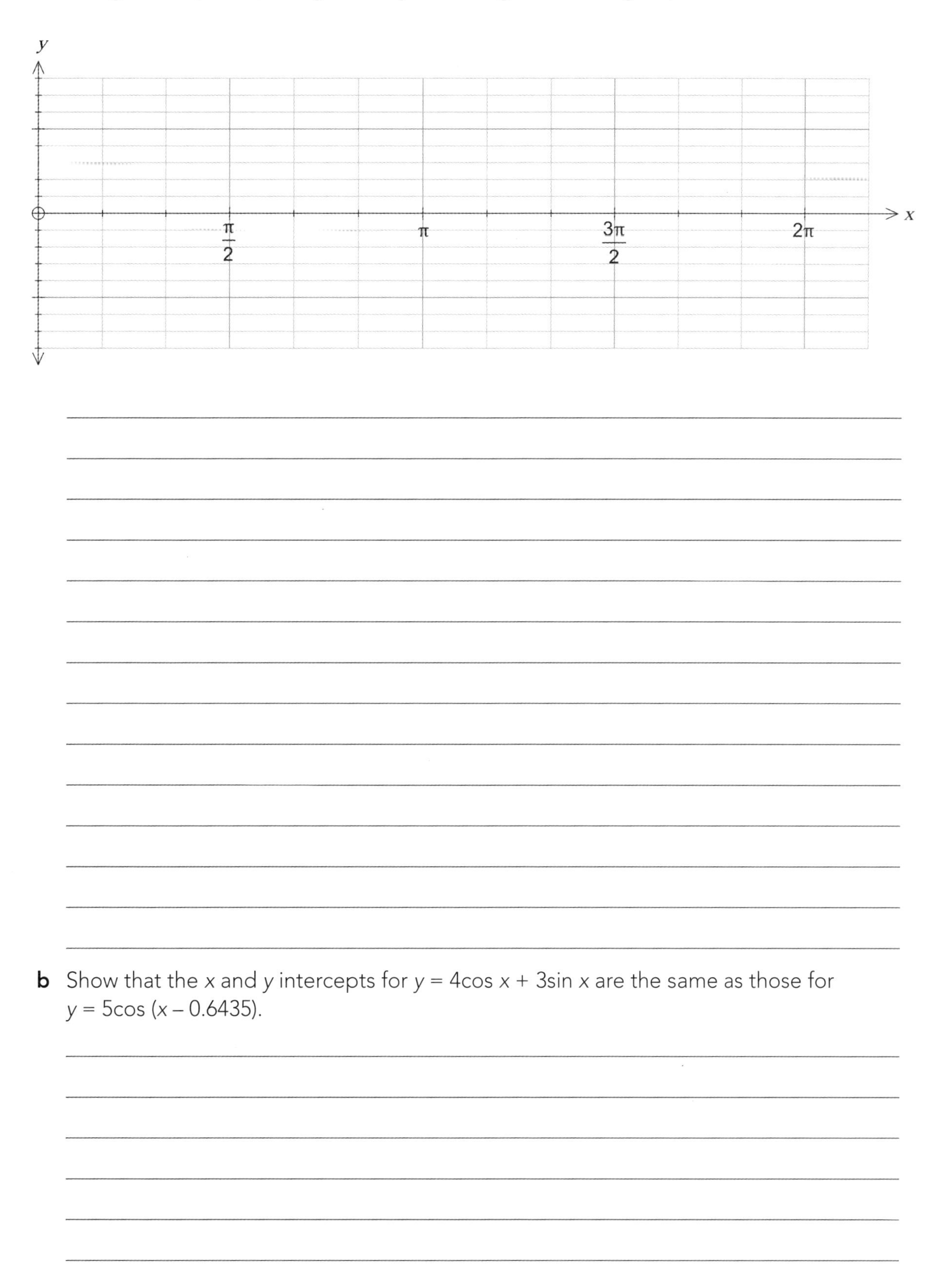

b Show that the x and y intercepts for $y = 4\cos x + 3\sin x$ are the same as those for $y = 5\cos(x - 0.6435)$.

c Use identities to show that $5\cos(x - 0.6435) = 4\cos x + 3\sin x$.

d Find general and particular solutions to $5\cos(x - 0.6435) = 2$ for the range $0 < x < 2\pi$.

e Show how general and particular solutions to $4\cos x + 3\sin x = 2$ for the range $0 < x < 2\pi$ could be found by squaring both sides of $4\cos x = 2 - 3\sin x$.

ISBN: 9780170425728

3 The graph shows the number hours of daylight in Dunedin (y) on each day (x) of a non-leap year (365 days). The marked points show number of hours of daylight on the shortest and longest days.

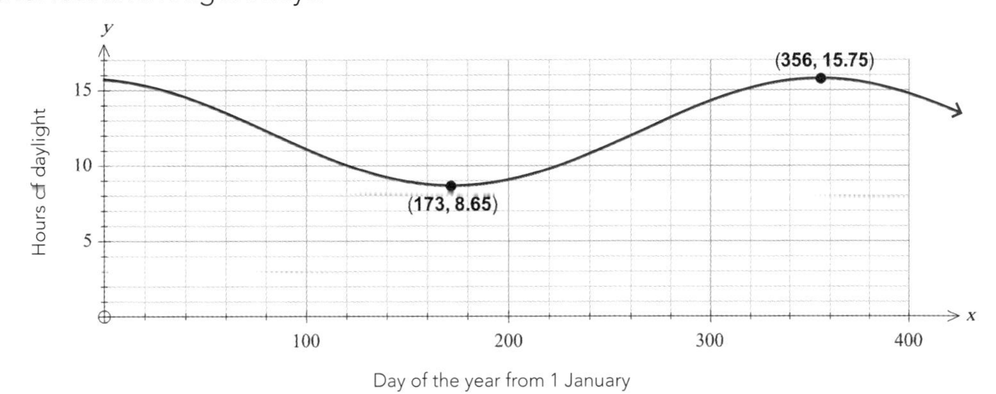

a Using the model $y = a\cos b(x \pm c) \pm d$, write the equation for this graph, showing how you found each value.

b On what days of the year in Dunedin are there 14 hours of daylight? Give the general solution and particular solutions for the first year.

c In Auckland, the shortest and longest days occur on the same days as in Dunedin. The shortest day has 9.6 hours of daylight, and the longest has 14.66 hours. Write an equation that models the day lengths for Auckland.

d Use your graphics calculator to find on what days of the year Dunedin and Auckland have the same number of daylight hours, and how much daylight they have on those days.

ISBN: 9780170425728

4 The picture shows a penny farthing, an old type of bicycle. The distance between the hub (centre of the front wheel) and the pedal is known as the crank. The crank for this penny farthing is 0.25 m long, and the radius of the front wheel is 0.7 m.

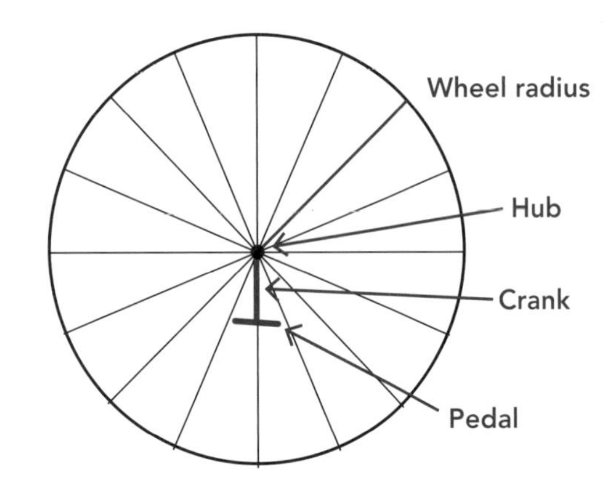

The diagram on the left shows detail of the front wheel.

The height of the pedal above the ground is modelled by the equation

$$h = 0.7 - 0.25\sin\frac{2\pi}{3}(t - 0.75)$$

a Draw the graph of $h = 0.7 - 0.25\sin\frac{2\pi}{3}(t - 0.75)$ for $0 \leq t \leq 6$ seconds. Let h represent the height of the pedal above the ground in metres, and t represent time in seconds. Show any reasoning and working needed to draw the graph.

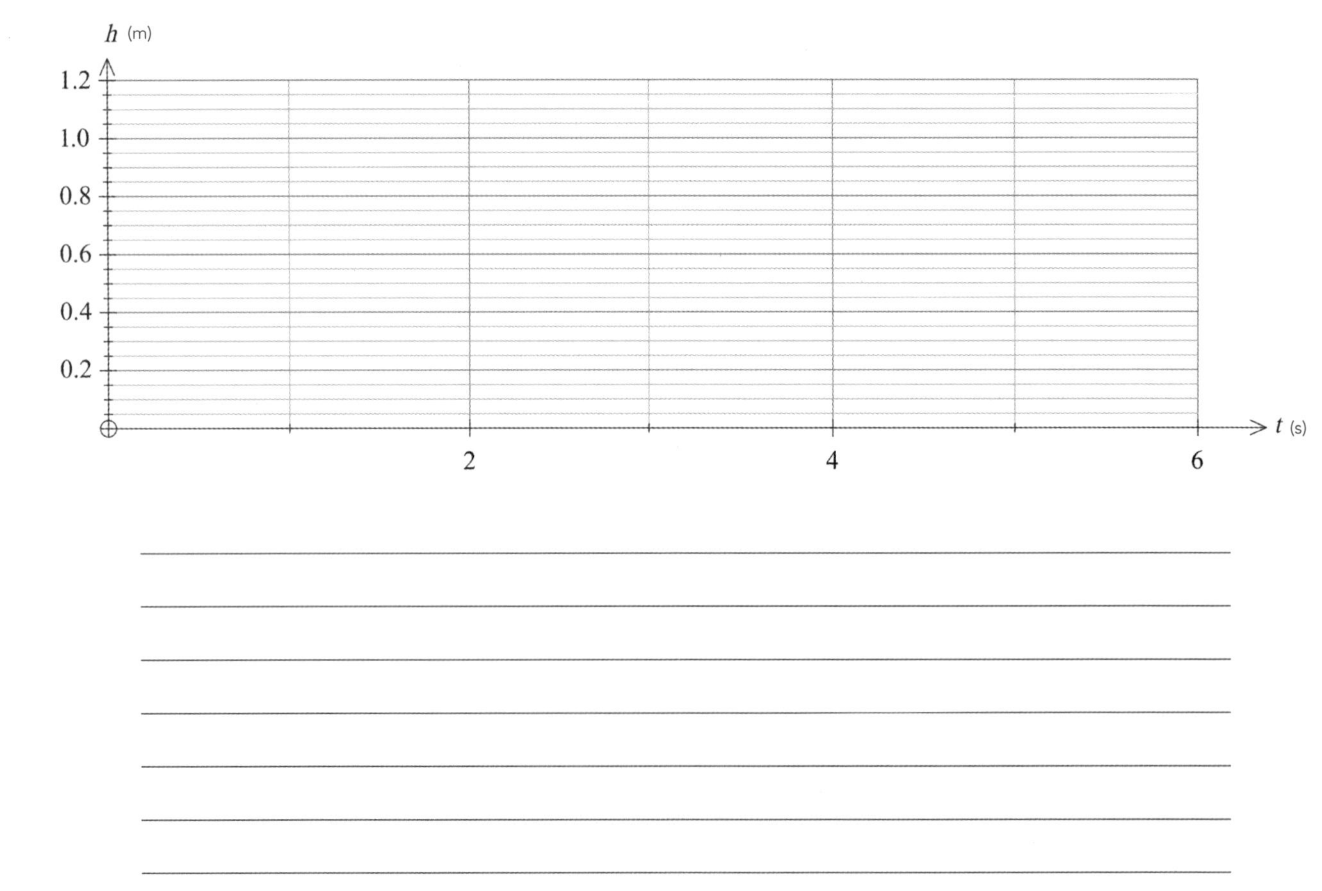

ISBN: 9780170425728

b How long does it take the cyclist to complete one revolution of the wheel?

c Calculate the periods during the first 6 seconds when the pedal is less than 60 cm above the ground.

d The cyclist altered the rate at which he was pedalling, so that he completed one revolution of the pedals in 2 seconds. Everything else remained the same. Rewrite the equation to show this.

e A more modern bicycle has a wheel with a radius of 30 cm, and a 20 cm crank. Its cyclist completes a full revolution of the pedal in 1.5 seconds. Using the model $h = \pm a \sin b(t \pm c) \pm d$, write an equivalent equation to that in **a** that models the height of the pedal above the ground. Where $t = 0$ seconds, the pedal should be at its maximum height above the ground. Show your reasoning and any calculations.

ISBN: 9780170425728

Practice tasks

Practice task one

Introduction
A design for a border is shown below:

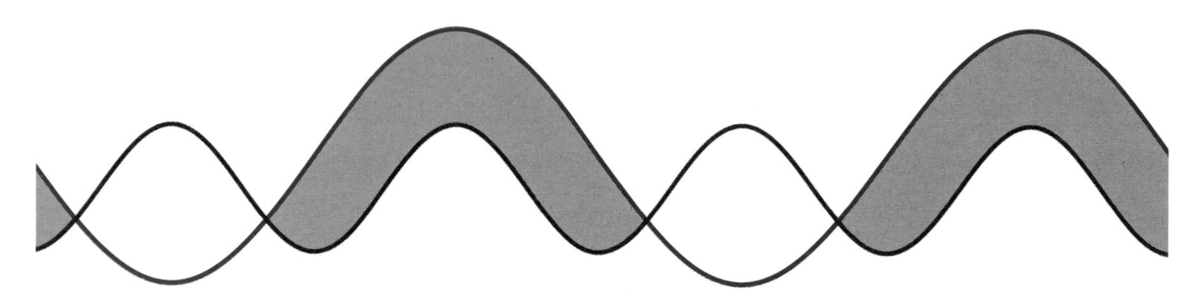

- The design is based on two intersecting curves: curve A (———) and curve B (———).
- Both curves are shown between 0 and 4π radians.

Task
This task requires you to use the information provided about the pattern to:
- Identify a mathematical model for curve A.
- Find a mathematical model for curve B.

The actual border will extend well beyond 4π, so you need to:
- Find the coordinates of the points for $0 < x < 4\pi$, between which the shading occurs.
- Find the general solution for the values between which the shading occurs.

Information:
Curve A has a y intercept at 0.5 and passes through the point $(4\pi, 0.5)$. A line between these points forms its axis of symmetry. It has an amplitude of 2. It can be modelled by one of the following equations:

$y = -2\sin x + 0.5$	$y = -2\sin 2x + 0.5$	$y = 2\sin 2x + 0.5$
$y = \sin 2x + 0.5$	$y = 2\sin x + 0.5$	$y = -\sin 2x + 0.5$

Curve B has a minimum at $(0, -1)$ and a maximum at $(\frac{\pi}{2}, 1)$ and is modelled on a cosine curve.

You must present your findings supported by any graphs and equations that you have used, along with any relevant calculations and links to the context. Communicate your methods clearly using appropriate mathematical statements.

ISBN: 9780170425728

Practice task two

Introduction
The graph shows the curve represented by the function $y = \cos 4\theta - 2\sin^2 2\theta$.

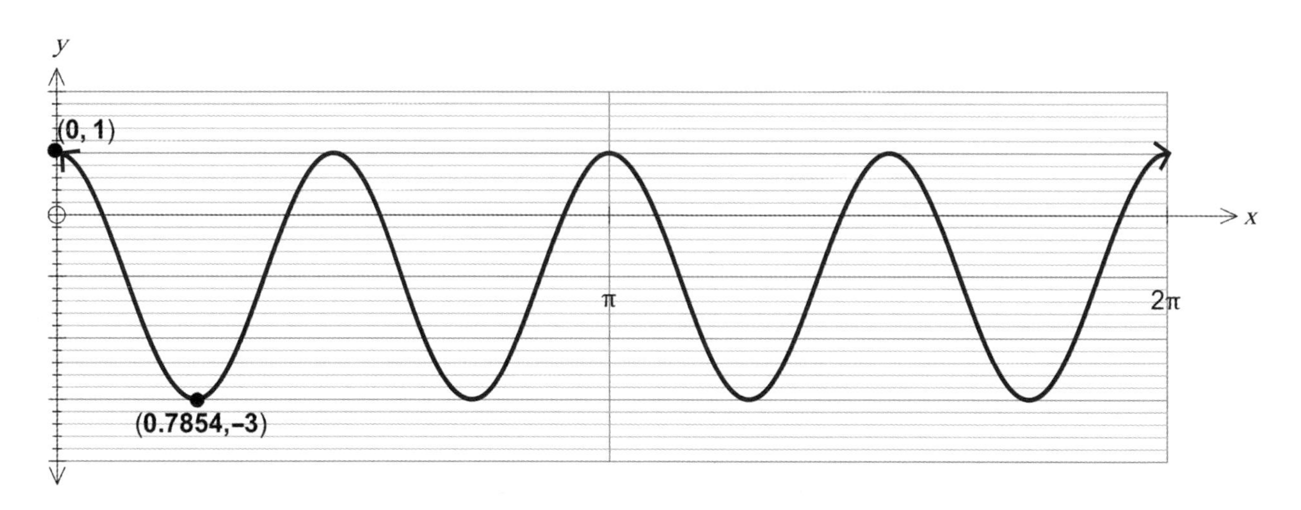

Fred maintains that the much simpler cosine function $y = 2\cos 4\theta - 1$ could be used to represent this curve.

Task
This task requires you to determine in as **many ways as you can**, whether or not Fred is correct. Your report could include:

- A discussion of graphical features (amplitude, periodicity, etc.).
- Comparison of the coordinates of graphical features (intercepts, maxima, minima, etc.).
- The use of trigonometrical identities and/or proofs.

You must present your findings supported by any graphs and equations that you have used, along with any relevant calculations and links to the context. Indicate any identities you have used. Communicate your methods clearly, using appropriate mathematical statements.

ISBN: 9780170425728

Practice task three

Introduction

A farmer near Whangarei plans to plant seeds for a crop of cucumbers, and he discovers that it is best to plant them when average temperatures are at least 16°C. He has been given an equation that models the mean monthly temperatures in the area, and a table of the actual mean temperatures. He needs help to understand this information, and advice about when he should plant.

Task

This task requires you to present the information to the farmer, along with advice about when it will be suitable to plant. Your report should include:

- A graph of the mathematical model for the monthly temperatures, along with a description of how you used the equation to find its significant features.
- A description of any periods of the year when the model is a poor fit for the actual data.
- An estimate, based on the information given, of the earliest date on which he should plant his seed.

Information

- Equation that models the mean monthly temperatures throughout the year:
  $T = 15.8 + 4.3\sin\frac{\pi}{6}(m + 1.65)$, where T represents temperature in °C and m represents months of the year, where January is month 1, February is month 2, etc.
- Actual mean monthly temperatures throughout the year:

Month	Jan	Feb	Mar	Apr	May	Jun	Jul	Aug	Sep	Oct	Nov	Dec
Temp (°C)	19.9	20.2	18.7	16.6	14.4	12.4	11.6	11.9	13.3	14.6	16.4	18.5

You must present your findings supported by any graphs and equations that you have used, along with any relevant calculations and links to the context. Communicate your methods clearly, using appropriate mathematical statements.

 ISBN: 9780170425728

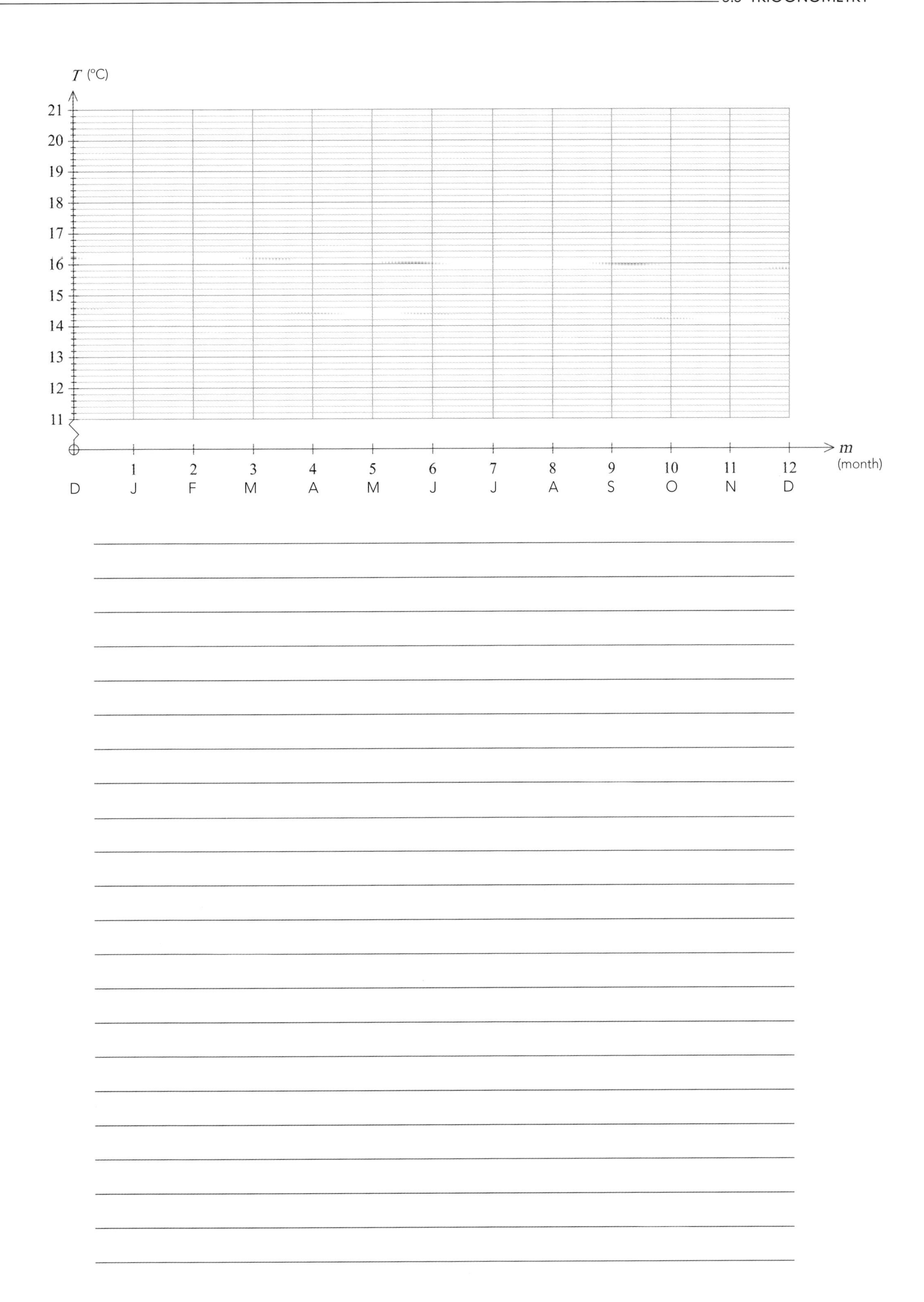

T (°C)
21
20
19
18
17
16
15
14
13
12
11
m (month)
1 2 3 4 5 6 7 8 9 10 11 12
D J F M A M J J A S O N D

Answers

Radians (pp. 6–8)

1 2π or 6.2832

2 $\frac{\pi}{2}$ or 1.5708

3 $\frac{\pi}{6}$ or 0.5236

4 $\frac{5\pi}{4}$ or 3.9270

5 3π or 9.4248

6 $\frac{\pi}{12}$ or 0.2618

7 0.4189

8 1.7279

9 1.0123

10 1.9897

11

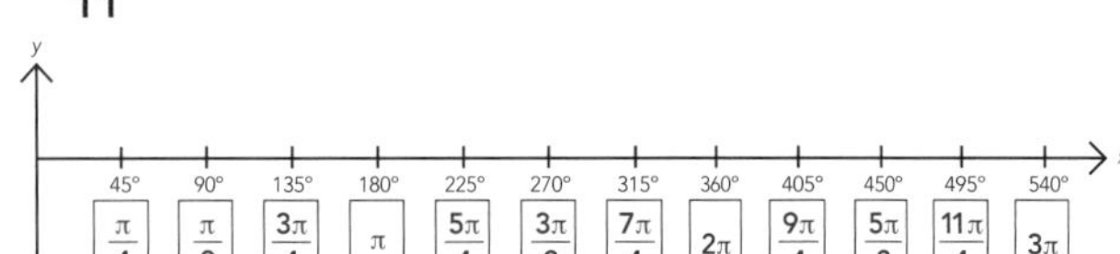

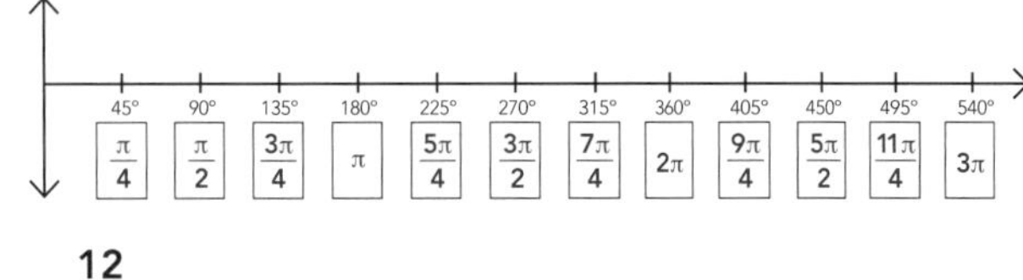

12

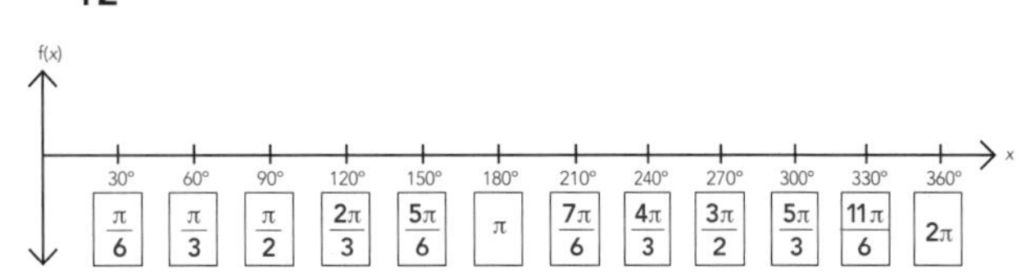

13 540° **14** 90° **15** 36°

16 54° **17** 17.2° **18** 70°

19 286.5° **20** 75° **21** 57.30°

22 255° **23** 0 **24** 1

25 Undefined **26** –1 **27** 0.5

28 –1 **29** 0.7071 **30** 0.7071

31 0.2588 **32** –0.8660 **33** 0.7648

34 –0.6878

Trigonometric functions (pp. 9–17)

The basic trigonometric functions (p. 11)

1 Period $= \frac{\pi}{2}$ Frequency = 4 Amplitude = 2

2 Period $= 4\pi$ Frequency = 0.5 Amplitude = 3

3 Period $= \pi$ Frequency = 2 Amplitude = 0.75

4 Period $= 2.5\pi$ Frequency = 0.8 Amplitude = 0.4

Another way of viewing trigonometric functions (pp. 13–17)

1 a $\theta = \frac{\pi}{3}$ or $\frac{5\pi}{3}$

b $\theta = \frac{\pi}{3} \Rightarrow \sin\theta = \frac{\sqrt{3}}{2}$ and $\tan\theta = \sqrt{3}$

$\theta = \frac{5\pi}{3} \Rightarrow \sin\theta = -\frac{\sqrt{3}}{2}$ and $\tan\theta = -\sqrt{3}$

2 a $\theta = \frac{\pi}{4}$ or $\frac{5\pi}{4}$

b $\theta = \frac{\pi}{4} \Rightarrow \sin\theta = \frac{1}{\sqrt{2}}$ and $\cos\theta = \frac{1}{\sqrt{2}}$

$\theta = \frac{5\pi}{4} \Rightarrow \sin\theta = -\frac{1}{\sqrt{2}}$ and $\cos\theta = -\frac{1}{\sqrt{2}}$

3 a $\theta = \frac{2\pi}{3}$ or $\frac{4\pi}{3}$

b $\theta = \frac{2\pi}{3} \Rightarrow \sin\theta = \frac{\sqrt{3}}{2}$ and $\tan\theta = -\sqrt{3}$

$\theta = \frac{4\pi}{3} \Rightarrow \sin\theta = -\frac{\sqrt{3}}{2}$ and $\tan\theta = \sqrt{3}$

4 a $\theta = \frac{5\pi}{6}$ or $\frac{11\pi}{6}$

b $\theta = \frac{5\pi}{6} \Rightarrow \sin\theta = -\frac{\sqrt{3}}{2}$ and $\cos\theta = \frac{\sqrt{3}}{2}$

$\theta = \frac{11\pi}{6} \Rightarrow \sin\theta = -\frac{1}{2}$ and $\cos\theta = \frac{\sqrt{3}}{2}$

5 $\cos\theta = \frac{4}{5} \Rightarrow \theta$ is in first quadrant or the fourth quadrant.

First quadrant $\Rightarrow \sin\theta = \frac{3}{5}$ and $\tan\theta = \frac{3}{4}$

Fourth quadrant $\Rightarrow \sin\theta = -\frac{3}{5}$ and $\tan\theta = -\frac{3}{4}$

6 $\sin\theta = -\frac{1}{3} \Rightarrow \theta$ is in third quadrant or the fourth quadrant.

Third quadrant $\Rightarrow \cos\theta = -\frac{\sqrt{8}}{3}$ and $\tan\theta = \frac{1}{\sqrt{8}}$

Fourth quadrant $\Rightarrow \cos\theta = \frac{\sqrt{8}}{3}$ and $\tan\theta = -\frac{1}{\sqrt{8}}$

7 $\tan\theta = \frac{3}{2} \Rightarrow \theta$ is in first quadrant or the third quadrant.

First quadrant $\Rightarrow \sin\theta = \frac{3}{\sqrt{13}}$ and $\cos\theta = \frac{2}{\sqrt{13}}$

Third quadrant $\Rightarrow \sin\theta = -\frac{3}{\sqrt{13}}$ and $\cos\theta = -\frac{2}{\sqrt{13}}$

8 $\tan\theta = \frac{5}{12} \Rightarrow \theta$ is in first quadrant or the second quadrant.

First quadrant $\Rightarrow \cos\theta = \frac{12}{13}$ and $\sin\theta = \frac{5}{13}$

Second quadrant $\Rightarrow \cos\theta = -\frac{12}{13}$ and $\sin\theta = -\frac{5}{13}$

9 $\cos\theta = -0.6$ or $-\frac{3}{5} \Rightarrow \theta$ is in second quadrant or the third quadrant.

Second quadrant $\Rightarrow \sin\theta = 0.8$ or $\frac{4}{5}$ and $\tan\theta = -1.3$ or $-\frac{4}{3}$

Third quadrant $\Rightarrow \sin\theta = -0.8$ or $-\frac{4}{5}$ and $\tan\theta = 1.3$ or $\frac{4}{3}$

ISBN: 9780170425728

Transformations of trigonometric functions (pp. 18–43)

Translations (pp. 18–28)

1 Vertical translations (pp. 19–20)

1

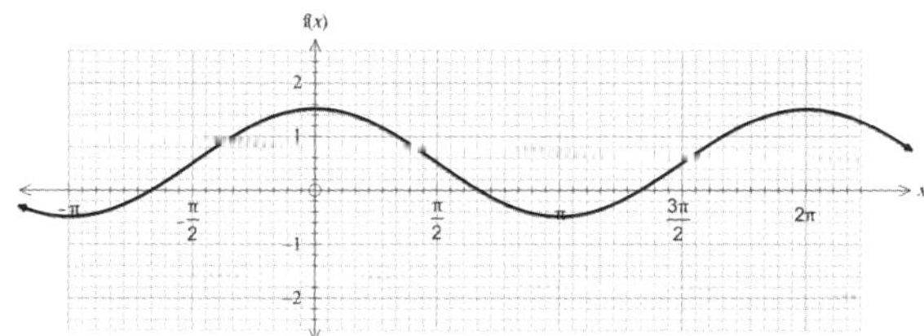

2

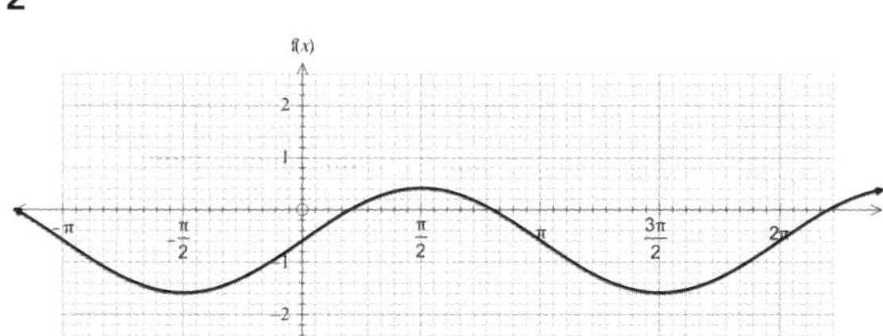

3

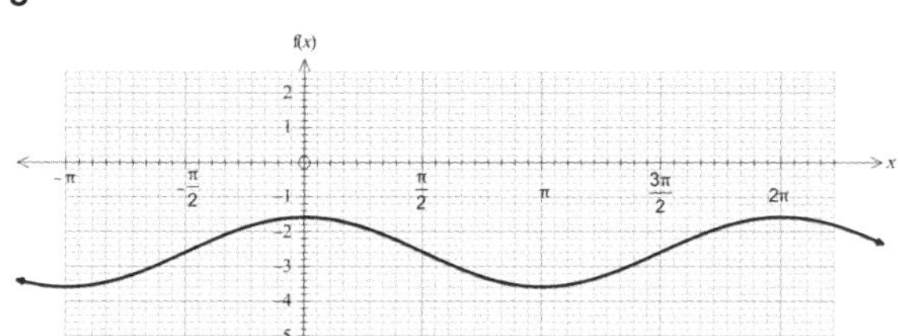

2 Horizontal translations (pp. 21–22)

1

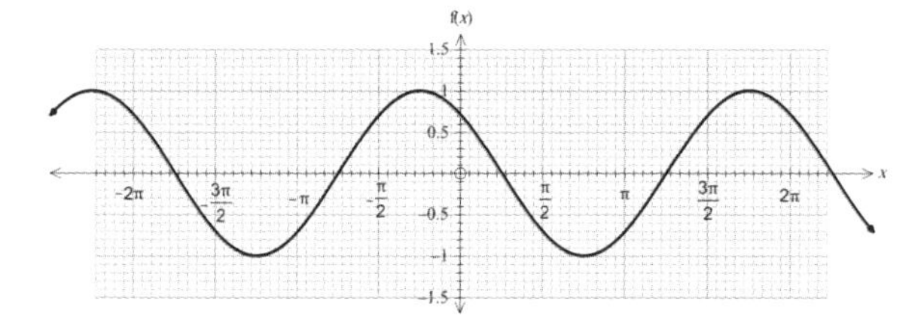

2

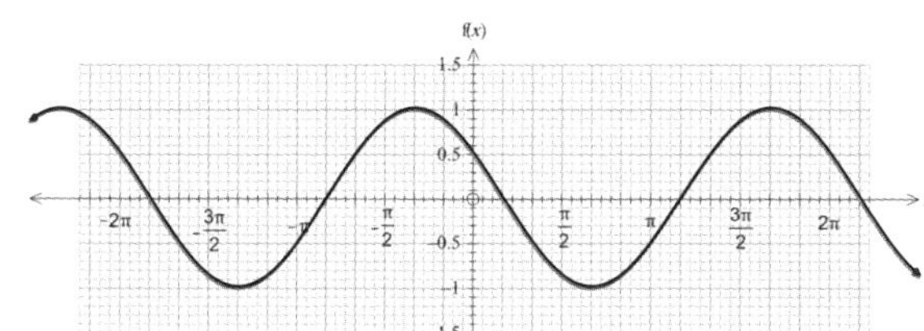

3

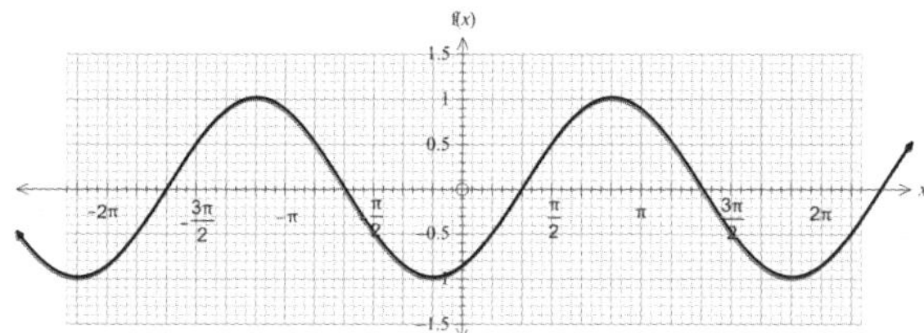

3 Combinations of translations (pp. 23–29)

1 C

2 A

3 D

4 B

5 $f(x) = 2.5\sin x + 1.5$

6 $f(x) = 5.3\cos x - 2.9$

7

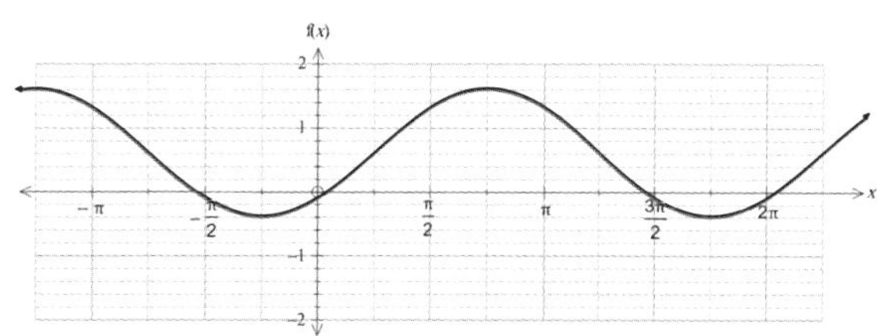

8

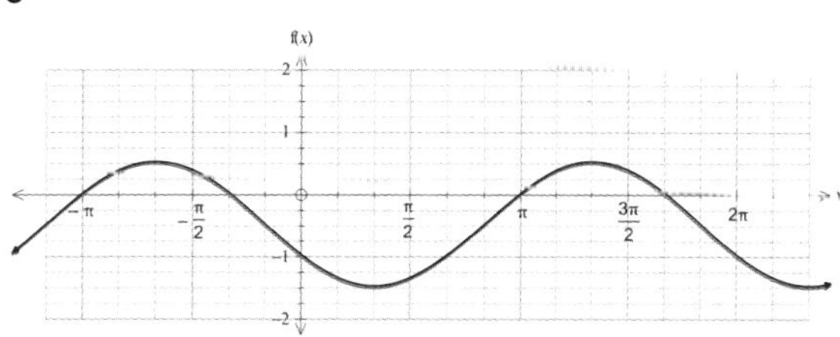

9 Any of:

$f(x) = \sin(x + \frac{3\pi}{4}) + 2.5$

$f(x) = \sin(x - \frac{5\pi}{4}) + 2.5$

$f(x) = \cos(x + \frac{\pi}{4}) + 2.5$

$f(x) = \cos(x - \frac{7\pi}{4}) + 2.5$

10 Any of:

$f(x) = \sin(x + \frac{4\pi}{3}) - 3.2$

$f(x) = \sin(x - \frac{2\pi}{3}) - 3.2$

$f(x) = \cos(x + \frac{5\pi}{6}) - 3.2$

$f(x) = \cos(x - \frac{7\pi}{6}) - 3.2$

11 Any of:

$f(x) = \sin(x + \frac{9\pi}{8}) - 1.6$

$f(x) = \sin(x - \frac{7\pi}{8}) - 1.6$

$f(x) = \cos(x + \frac{5\pi}{8}) - 1.6$

$f(x) = \cos(x - \frac{11\pi}{8}) - 1.6$

Enlargements (pp. 30–35)

1 Vertical englargements (pp. 30–31)

1

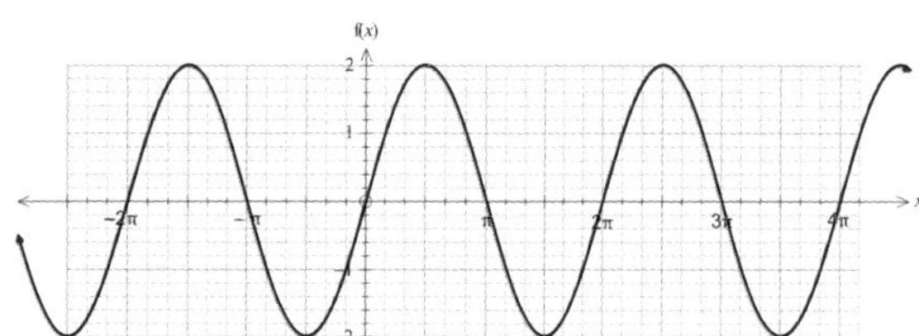

2

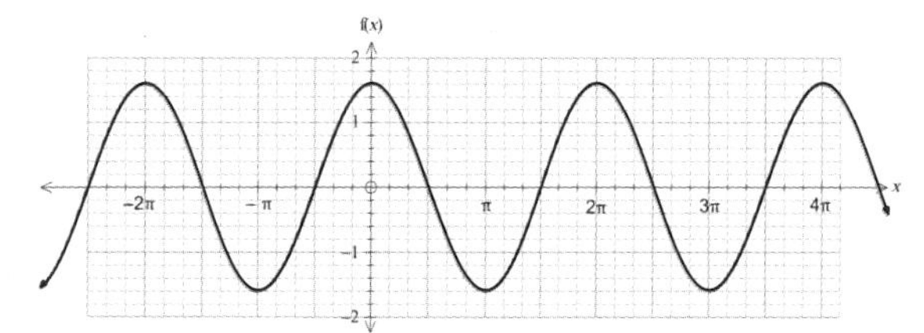

2 Horizontal enlargements (pp. 32–33)

1

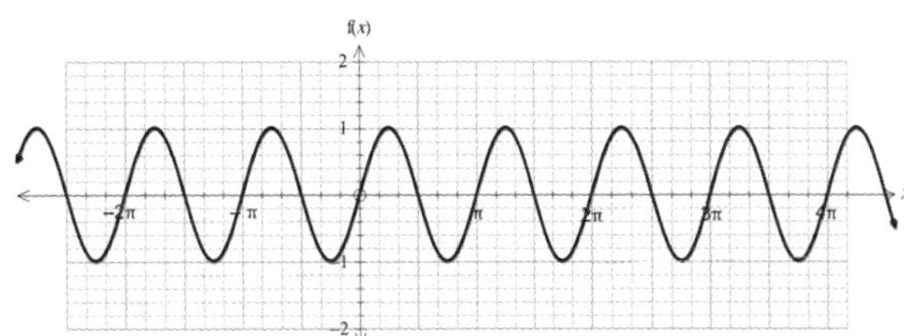

2

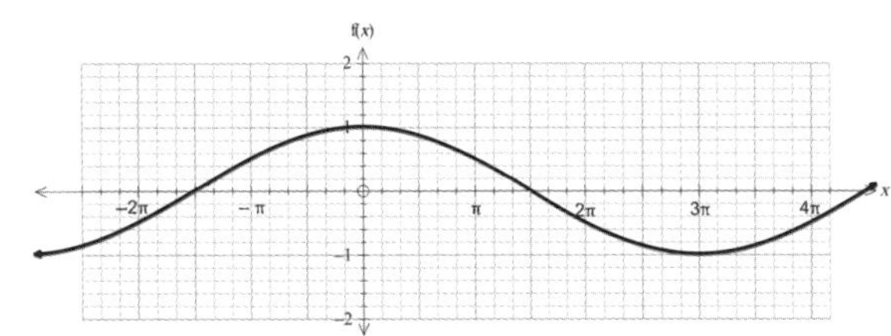

3 Combinations of enlargements (pp. 33–35)

1 D **2** A **3** E **4** B

5 C **6** $f(x) = 1.6\cos 4x$

7 $f(x) = 0.4\cos \frac{1}{3}x$ **8** $f(x) = 0.75\sin 3x$

Putting it all together (pp. 36–43)

1 C **2** B **3** D **4** A

5 $f(x) = 1.2\sin 3(x + \frac{\pi}{6}) + 0.8$ or

$f(x) = 1.2\sin 3(x - \frac{\pi}{2}) + 0.8$

6 $f(x) = 0.4\cos 0.5(x + \frac{13\pi}{4}) - 0.6$ or

$f(x) = 0.4\cos 0.5(x - \frac{3\pi}{4}) - 0.6$

7 $f(x) = 1.5\sin 4(x + \frac{3\pi}{8}) + 3.5$ or

$f(x) = 1.5\sin 4(x - \frac{\pi}{8}) + 3.5$

8

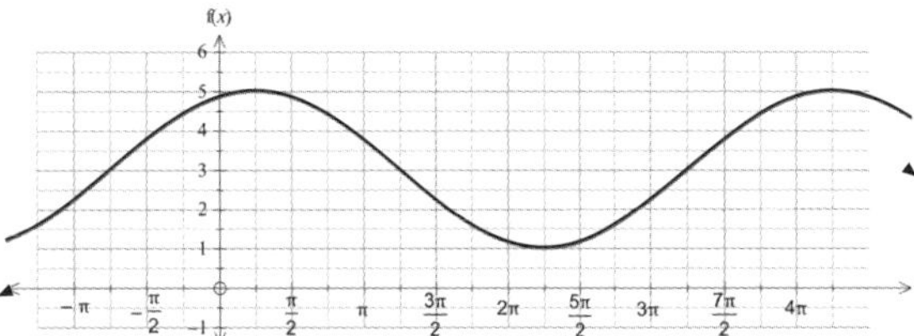

9

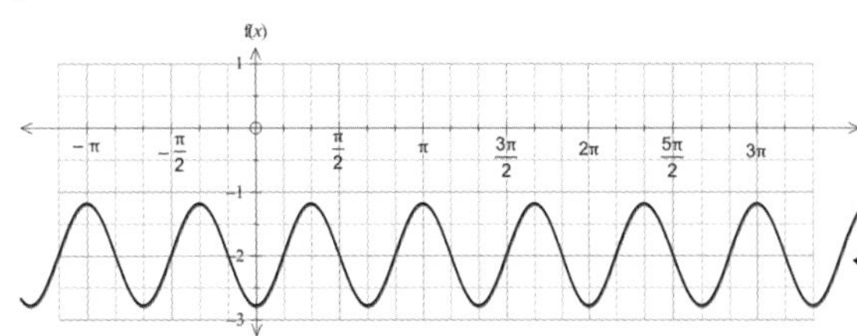

10

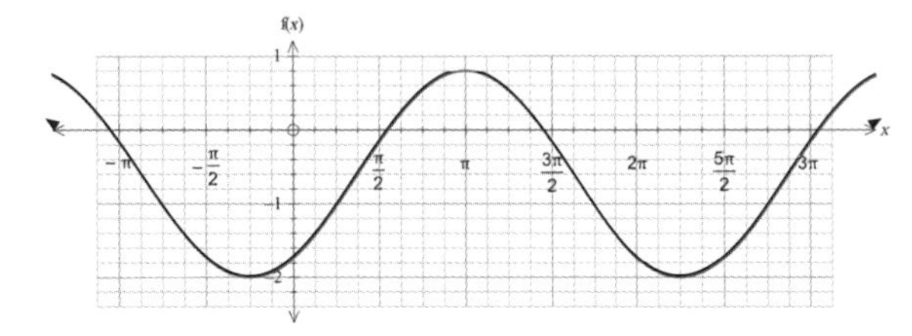

11

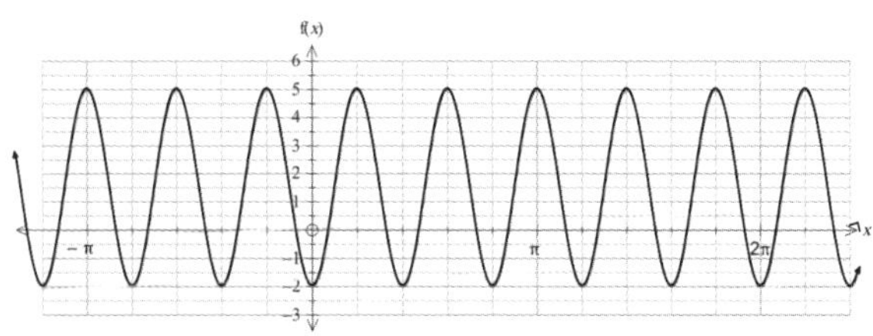

Trigonometric identities (pp. 46–90)

Reciprocal trigonometric functions and identities (pp. 46–47)

1 1 **2** −0.4577

3 $0.\dot{3}$ **4** −5.1713

5 $6 + 1 = 7$ **6** $-3.3507 + 4.2426 = 0.8919$

Two more trigonometric identities and simple proofs (pp. 49–52)

1 $\cot\theta \times \sin\theta = \frac{\cos\theta}{\sin\theta} \times \frac{\sin\theta}{1}$

$= \cos\theta$

2 $\tan\theta \times \cos\theta \times \operatorname{cosec}\theta = \frac{\sin\theta}{\cos\theta} \times \frac{\cos\theta}{1} \times \frac{1}{\sin\theta}$

$= 1$

3 $\frac{2\cos\theta}{\cot\theta} = \frac{2\cos\theta}{\frac{\cos\theta}{\sin\theta}}$

$= \frac{2\cos\theta}{1} \times \frac{\sin\theta}{\cos\theta}$

$= 2\sin\theta$

4 $\sec^2\theta \times \sin\theta \times \cot^2\theta$

$= \frac{1}{\cos^2\theta} \times \frac{\sin\theta}{1} \times \frac{\cos^2\theta}{\sin^2\theta} = \frac{1}{\sin\theta}$

$= \operatorname{cosec}\theta$

5 $\frac{1 + \cos\theta}{\sin\theta} = \frac{1}{\sin\theta} + \frac{\cos\theta}{\sin\theta}$

$= \operatorname{cosec}\theta + \cot\theta$

6 $\frac{\sin\theta + 3}{\cos\theta} = \frac{\sin\theta}{\cos\theta} + \frac{3}{\cos\theta}$

$= \tan\theta + 3\sec\theta$

7 $\frac{\sin\theta - \sec\theta}{\tan\theta}$

$= \frac{\sin\theta}{\frac{\sin\theta}{\cos\theta}} - \frac{\frac{1}{\cos\theta}}{\frac{\sin\theta}{\cos\theta}}$

$= \frac{\sin\theta}{1} \times \frac{\cos\theta}{\sin\theta} - \frac{1}{\cos\theta} \times \frac{\cos\theta}{\sin\theta}$

$= \cos\theta - \frac{1}{\sin\theta}$

$= \cos\theta - \operatorname{cosec}\theta$

8 $\tan\theta \times \cos^2\theta - \cot\theta \times \sin^2\theta$

$= \frac{\sin\theta}{\cos\theta} \times \frac{\cos^2\theta}{1} - \frac{\cos\theta}{\sin\theta} \times \frac{\sin^2\theta}{1}$

$= \sin\theta \times \cos\theta - \cos\theta \times \sin\theta$

$= 0$

ISBN: 9780170425728

9 $\dfrac{3-\tan\theta}{\sin\theta} - 2\operatorname{cosec}\theta$

$= \dfrac{3-\tan\theta}{\sin\theta} - \dfrac{2}{\sin\theta}$

$= \dfrac{3-\tan\theta-2}{\sin\theta}$

$= \dfrac{1-\tan\theta}{\sin\theta}$

$= \dfrac{1}{\sin\theta} - \left(\dfrac{\frac{\sin\theta}{\cos\theta}}{\frac{\sin\theta}{1}}\right)$

$= \dfrac{1}{\sin\theta} - \dfrac{\sin\theta}{\cos\theta} \times \dfrac{1}{\sin\theta}$

$= \operatorname{cosec}\theta - \sec\theta$

10 $\dfrac{3}{\cos\theta \times \operatorname{cosec}\theta} - \dfrac{2\sin\theta \times \operatorname{cosec}\theta}{\cot\theta}$

$= \dfrac{3}{\cos\theta \times \frac{1}{\sin\theta}} - \dfrac{2\sin\theta \times \frac{1}{\sin\theta}}{\cot\theta}$

$= \dfrac{3}{\cot\theta} - \dfrac{2}{\cot\theta}$

$= \tan\theta$

The Pythagorean identities (pp. 53–56)

1 $\sin^2\theta + \tan^2\theta = (1-\cos^2\theta) + (\sec^2\theta - 1)$
$= \sec^2\theta - \cos^2\theta$

2 $\cos\theta\,(1-\sin^2\theta) = \cos\theta \times \cos^2\theta$
$= \cos^3\theta$

3 $6\operatorname{cosec}^2\theta - 2(\cot^2\theta + 3) = 6(\cot^2\theta + 1) - 2(\cot^2\theta + 3)$
$= 6\cot^2\theta + 6 - 2\cot^2\theta - 6$
$= 4\cot^2\theta$

4 $\operatorname{cosec}^4\theta - \cot^4\theta - 1$
$= (\operatorname{cosec}^2\theta - \cot^2\theta)(\operatorname{cosec}^2\theta + \cot^2\theta) - 1$
$= ((\cot^2\theta + 1) - \cot^2\theta)((\cot^2\theta + 1) + \cot^2\theta) - 1$
$= (1)(2\cot^2\theta + 1) - 1$
$= 2\cot^2\theta$

5 $\dfrac{\cos\theta + \sin\theta}{\sin\theta} - 1 = \dfrac{\cos\theta + \sin\theta}{\sin\theta} - \dfrac{\sin\theta}{\sin\theta}$

$= \dfrac{\cos\theta + \sin\theta - \sin\theta}{\sin\theta}$

$= \dfrac{\cos\theta}{\sin\theta}$

6 $\dfrac{\cos A}{1+\sin A} = \dfrac{\cos A}{1+\sin A} \times \dfrac{1-\sin A}{1-\sin A}$

$= \dfrac{\cos A(1-\sin A)}{1-\sin^2 A}$

$= \dfrac{\cos A(1-\sin A)}{\cos^2 A}$

$= \dfrac{1-\sin A}{\cos A}$

7 $\dfrac{\cos\theta}{1+\sin\theta} - \dfrac{1-\sin\theta}{\cos\theta} = \dfrac{\cos^2\theta - (1-\sin\theta)(1+\sin\theta)}{\cos\theta\,(1+\sin\theta)}$

$= \dfrac{\cos^2\theta - 1 + \sin^2\theta}{\cos\theta\,(1+\sin\theta)}$

$= \dfrac{1-1}{\cos\theta\,(1+\sin\theta)}$

$= 0$

Mixing it up (pp. 57–60)

1 $\dfrac{1}{1+\sin\theta} + \dfrac{1}{1-\sin\theta} = \dfrac{1-\sin\theta + 1 + \sin\theta}{(1+\sin\theta)(1-\sin\theta)}$

$= \dfrac{2}{1-\sin^2\theta}$

$= \dfrac{2}{\cos^2\theta}$

$= 2\sec^2\theta$

2 $(\cot^2\theta + 1)(\sec^2\theta - 1) = \operatorname{cosec}^2\theta \times \tan^2\theta$

$= \dfrac{1}{\sin^2\theta} \times \dfrac{\sin^2\theta}{\cos^2\theta}$

$= \dfrac{1}{\cos^2\theta}$

$= \sec^2\theta$

3 $(1-\sin\theta)(1+\operatorname{cosec}\theta)$
$= 1 + \operatorname{cosec}\theta - \sin\theta - \sin\theta\operatorname{cosec}\theta$
$= \operatorname{cosec}\theta - \sin\theta$

$= \dfrac{1}{\sin\theta} - \sin\theta$

$= \dfrac{1-\sin^2\theta}{\sin\theta}$

$= \dfrac{\cos^2\theta}{\sin\theta}$

$= \cos\theta\cot\theta$

4 $\dfrac{\tan^2\theta + 1}{\sec\theta} = \dfrac{\sec^2\theta}{\sec\theta}$

$= \sec\theta$

$= \dfrac{1}{\cos\theta}$

$= \dfrac{\cos^2\theta}{\cos^3\theta}$

$= \dfrac{1-\sin^2\theta}{\cos^3\theta}$

5 $\dfrac{\cos^4\theta - \sin^4\theta}{\sin^2\theta} = \dfrac{(\cos^2\theta - \sin^2\theta)(\cos^2\theta + \sin^2\theta)}{\sin^2\theta}$

$= \dfrac{(\cos^2\theta - \sin^2\theta)(1)}{\sin^2\theta}$

$= \dfrac{\cos^2\theta}{\sin^2\theta} - \dfrac{\sin^2\theta}{\sin^2\theta}$

$= \cot^2\theta - 1$

6 $\frac{1 + \sin \theta}{\cos \theta} = \frac{1 + \sin \theta}{\cos \theta} \times \frac{1 - \sin \theta}{1 - \sin \theta}$

$= \frac{1 - \sin^2 \theta}{\cos \theta (1 - \sin \theta)}$

$= \frac{\cos^2 \theta}{\cos \theta(1 - \sin \theta)}$

$= \frac{\cos \theta}{1 - \sin \theta}$

7 $\cos \theta - \cot^2 \theta \times \sec \theta = \cos \theta - \frac{\cos^2 \theta}{\sin^2 \theta} \times \frac{1}{\cos \theta}$

$= \cos \theta - \frac{\cos \theta}{\sin^2 \theta}$

$= \cos \theta \left(1 - \frac{1}{\sin^2 \theta}\right)$

$= \cos \theta(1 - \operatorname{cosec}^2 \theta)$

$= \cos \theta \times \tan^2 \theta$

$= \cos \theta \times \frac{\sin^2 \theta}{\cos^2 \theta}$

$= \frac{\sin^2 \theta}{\cos \theta}$

$= \sin \theta \tan \theta$

8 $\frac{\sec \theta}{\tan \theta - 1} - \frac{\tan \theta - 1}{\sec \theta} = \frac{\sec^2 \theta - (\tan \theta - 1)^2}{\sec \theta (\tan \theta - 1)}$

$= \frac{\sec^2 \theta - \tan^2 \theta + 2\tan \theta - 1}{\sec \theta (\tan \theta - 1)}$

$= \frac{\tan^2 \theta - \tan^2 \theta + 2\tan \theta}{\sec \theta (\tan \theta - 1)}$

$= \frac{2 \tan \theta}{\sec \theta (\tan \theta - 1)}$

$= \frac{\frac{2\sin \theta}{\cos \theta}}{\frac{\tan \theta - 1}{\cos \theta}}$

$= \frac{2\sin \theta}{\tan \theta - 1}$

9 $\frac{\sec \theta}{\sec \theta - \cos \theta} = \frac{\frac{1}{\cos \theta}}{\frac{1}{\cos \theta} - \frac{\cos^2 \theta}{\cos \theta}}$

$= \frac{\frac{1}{\cos \theta}}{\frac{1 - \cos^2 \theta}{\cos \theta}}$

$= \frac{1}{\cos \theta} \times \frac{\cos \theta}{\sin^2 \theta}$

$= \operatorname{cosec}^2 \theta$

10 $\frac{\sec \theta}{\sin \theta + 1} = \frac{\frac{1}{\cos \theta}}{\sin \theta + 1} \times \frac{\sin \theta - 1}{\sin \theta - 1}$

$= \frac{\frac{\sin \theta - 1}{\cos \theta}}{\sin^2 \theta - 1}$

$= \frac{\sin \theta - 1}{\cos \theta} \times \frac{1}{-\cos^2 \theta}$

$= \frac{\sin \theta - 1}{-\cos^3 \theta}$

$= \frac{1 - \sin \theta}{\cos^3 \theta}$

Angle sums and differences (pp. 61–68)

1 Simplifying expressions containing angle sums and differences (pp. xx)

1 $\sin A$ **2** $-\sin a$
3 $\tan b$ **4** $\operatorname{cosec} A$
5 $-\sec b$ **6** $\cot B$

2 Finding exact values of angles sums and differences (pp. xx)

1 $\frac{\sqrt{3} + 1}{2\sqrt{2}}$ **2** $\frac{1 + \sqrt{3}}{1 - \sqrt{3}}$
3 $\frac{\sqrt{3} + 1}{2\sqrt{2}}$ **4** $\frac{2\sqrt{2}}{1 - \sqrt{3}}$
5 $-\sqrt{2}$ **6** $\frac{2}{\sqrt{3}}$

3 Proving identities with expressions containing angle sums and differences (pp. XX)

1 $\cos (A + B) - \cos (A - B)$
$= (\cos A \cos B - \sin A \sin B) - (\cos A \cos B + \sin A \sin B)$
$= \cos A \cos B - \sin A \sin B - \cos A \cos B - \sin A \sin B$
$= -2\sin A \sin B$

2 $\sin (A + B) \sin (A - B) = (\sin A \cos B + \cos A \sin B)(\sin A \cos B - \cos A \sin B)$

$= \sin^2 A \cos^2 B - \cos^2 A \sin^2 B$
(difference of two squares)

$= \sin^2 A(1 - \sin^2 B) - (1 - \sin^2 A)\sin^2 B$
(LHS contains only sines)

$= \sin^2 A - \sin^2 A \sin^2 B - \sin^2 B + \sin^2 A \sin^2 B$

$= \sin^2 A - \sin^2 B$

3 $\frac{\sin (x - y)}{\cos x \cos y} = \frac{\sin x \cos y - \cos x \sin y}{\cos x \cos y}$

$= \frac{\sin x \cos y}{\cos x \cos y} - \frac{\cos x \sin y}{\cos x \cos y}$

$= \frac{\sin x}{\cos x} - \frac{\sin y}{\cos y}$

$= \tan x - \tan y$

ISBN: 9780170425728

4 $\sin\left(\frac{\pi}{4} + A\right) + \sin\left(\frac{\pi}{4} - A\right)$

$= \sin\frac{\pi}{4}\cos A + \cos\frac{\pi}{4}\sin A$

$+ \sin\frac{\pi}{4}\cos A - \cos\frac{\pi}{4}\sin A$

$= \frac{1}{\sqrt{2}}\cos A + \frac{1}{\sqrt{2}}\cos A$

$= \frac{2}{\sqrt{2}}\cos A$

$- \sqrt{2}\cos A$

5 $\dfrac{\cos\left(\frac{\pi}{4} - \theta\right)}{\cos\frac{\pi}{4}\cos\theta} - \tan\theta$

$= \dfrac{\cos\frac{\pi}{4}\cos\theta + \sin\frac{\pi}{4}\sin\theta}{\cos\frac{\pi}{4}\cos\theta} - \dfrac{\sin\theta}{\cos\theta}$

$= \dfrac{\frac{1}{\sqrt{2}}\cos\theta + \frac{1}{\sqrt{2}}\sin\theta}{\frac{1}{\sqrt{2}}\cos\theta} - \dfrac{\sin\theta}{\cos\theta}$

$= \dfrac{\cos\theta + \sin\theta}{\cos\theta} - \dfrac{\sin\theta}{\cos\theta}$

$= \dfrac{\cos\theta + \sin\theta - \sin\theta}{\cos\theta}$

$= 1$

6 $\tan(\pi + \theta) = \dfrac{\tan\pi + \tan\theta}{1 - \tan\pi\tan\theta}$

$= \dfrac{0 + \tan\theta}{1 - (0)\tan\theta}$

$= \tan\theta$

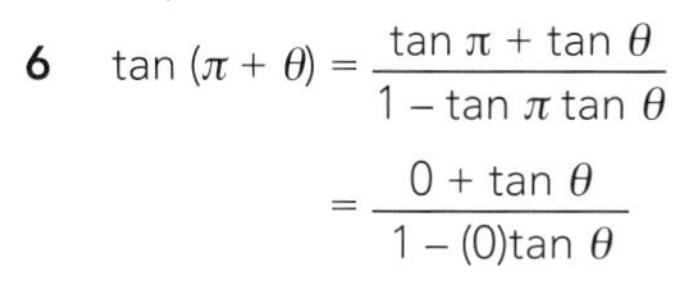

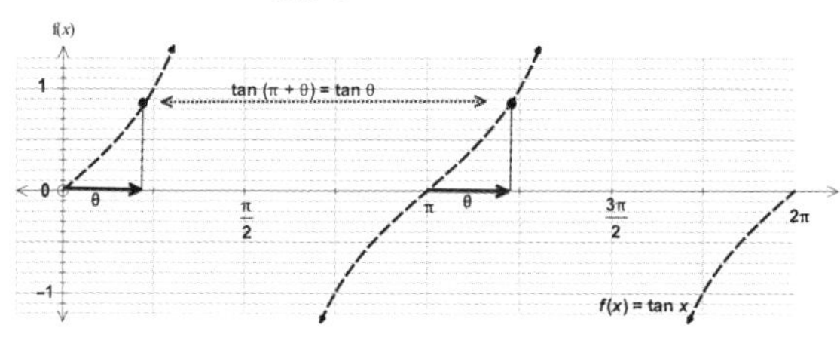

7 $\sin(\pi - \theta) = \sin\pi\cos\theta - \cos\pi\sin\theta$

$= 0 \times \cos\theta - (-1) \times \sin\theta$

$= \sin\theta$

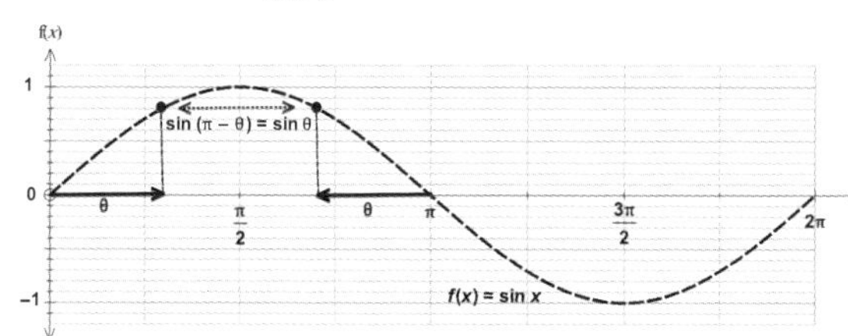

8 $\cos\left(\frac{\pi}{2} + \theta\right) = \cos\frac{\pi}{2}\cos\theta - \sin\frac{\pi}{2}\sin\theta$

$= 0 \times \cos\theta - 1 \times \sin\theta$

$= -\sin\theta$

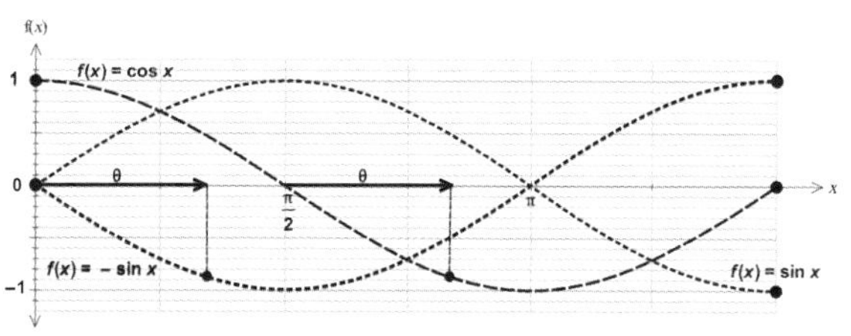

4 Bits and pieces (pp. xx)

1 $\sqrt{3}$

2 0

3 $\frac{6}{7}$

4 $\frac{6}{\sqrt{2}}\sin A = \frac{6}{\sqrt{2}}\cos A$ so $A = \frac{\pi}{4}$

$\therefore \tan A = 1$

Double angles (pp. 69–78)

1 Using the formulae (pp. 70)

1 sin 120° **2** $\cos\pi$

3 tan 2A **4** $-\cos\frac{\pi}{4}$

5 $\frac{1}{2}\sin 6$ **6** $-\cos\frac{2\pi}{3}$

2 Finding exact values of double angles (pp. 71–73)

1 **a** $\frac{3}{5}$ **b** $\frac{4}{5}$ **c** $\frac{3}{4}$

2 **a** $-\frac{\sqrt{3}}{2}$ **b** $-\frac{1}{2}$ **c** $\sqrt{3}$

3 **a** $\frac{24}{25}$

b $-\frac{7}{25}$

(Note the negative answer. While θ is in the first quadrant, 2θ lies in the second quadrant.)

c $-\frac{24}{7}$

(Again, the answer is negative for the same reason as **b**.)

4 **a** $\frac{3}{5}$ **b** $-\frac{4}{5}$ **c** $-\frac{3}{4}$

3 Using the formula to find half angles (pp. 74–76)

1 $\frac{2}{\sqrt{5}}$ **2** $\frac{1}{\sqrt{26}}$ **3** $\sqrt{\frac{\sqrt{5} + 5}{10}}$

4 $\frac{\sqrt{3}}{2}$

5 Using quadratic formula: $\tan\frac{\pi}{8} = \frac{-2 + \sqrt{8}}{2}$. Because $\frac{\pi}{8}$ is in the first quadrant, it must be positive $\therefore \tan\frac{\pi}{8} = \frac{-2 + \sqrt{8}}{2}$

6 2 (Not $-\frac{1}{2}$ because $\frac{\theta}{2}$ must be in the first quadrant where tangents are positive.)

4 Proving identities with expressions containing double angles (pp. 77–78)

1 $\dfrac{2}{1+\cos 2\theta} = \dfrac{2}{1+(2\cos^2\theta - 1)}$

$= \dfrac{2}{2\cos^2\theta}$

$= \sec^2\theta$

2 $\dfrac{1+\cos 2\theta}{1-\cos^2\theta}$

$= \dfrac{1+(2\cos^2\theta - 1)}{\sin^2\theta}$

$= \dfrac{2\cos^2\theta}{\sin^2\theta}$

$= 2\cot^2\theta$

3 $\dfrac{1-\tan^2\theta}{1+\tan^2\theta} = \dfrac{1-\tan^2\theta}{\sec^2\theta}$

$= \dfrac{1}{\sec^2\theta} - \dfrac{\tan^2\theta}{\sec^2\theta}$

$= \cos^2\theta - \dfrac{\sin^2\theta}{\cos^2\theta} \times \dfrac{\cos^2\theta}{1}$

$= \cos^2\theta - \sin^2\theta$

$= \cos 2\theta$

4 $\cos 4\theta = 2\cos^2 2\theta - 1$

$= 2(2\cos^2\theta - 1)^2 - 1$

$= 2(4\cos^4\theta - 4\cos^2\theta + 1) - 1$

$= 8\cos^4\theta - 8\cos^2\theta + 1$

5 $\dfrac{\sin\theta + \sin 2\theta}{1+\cos\theta + \cos 2\theta} = \dfrac{\sin\theta + 2\sin\theta\cos\theta}{1+\cos\theta + (2\cos^2\theta - 1)}$

$= \dfrac{\sin\theta\,(1+2\cos\theta)}{\cos\theta\,(1+2\cos\theta)}$

$= \tan\theta$

6 $2\cot 2\theta = \dfrac{2}{\tan 2\theta}$

$= \dfrac{2}{\dfrac{2\tan\theta}{1-\tan^2\theta}}$

$= \dfrac{2-2\tan^2\theta}{2\tan\theta}$

$= \dfrac{1}{\tan\theta} - \dfrac{\tan^2\theta}{\tan\theta}$

$= \cot\theta - \tan\theta$

7 $\dfrac{\cos\theta + \sin\theta}{\cos\theta - \sin\theta} = \dfrac{\cos\theta + \sin\theta}{\cos\theta - \sin\theta} \times \dfrac{\cos\theta + \sin\theta}{\cos\theta + \sin\theta}$

$= \dfrac{\cos^2\theta + 2\cos\theta\sin\theta + \sin^2\theta}{\cos^2\theta - \sin^2\theta}$

$= \dfrac{1+2\cos\theta\sin\theta}{\cos 2\theta}$

$= \dfrac{1+\sin 2\theta}{\cos 2\theta}$

8 $\cos 3\theta = \cos(2\theta + \theta)$

$= \cos 2\theta\cos\theta - \sin 2\theta\sin\theta$

$= (2\cos^2\theta - 1)\cos\theta - (2\cos\theta\sin\theta)\sin\theta$

$= 2\cos^3\theta - \cos\theta - 2\sin^2\theta\cos\theta$

$= 2\cos^3\theta - \cos\theta - 2(1-\cos^2\theta)\cos\theta$

$= 2\cos^3\theta - \cos\theta - 2\cos\theta + 2\cos^3\theta$

$= 4\cos^3\theta - 3\cos\theta$

Sums and products (pp. 79–90)

Products to sums or differences (pp. 79–82)

1 $\sin 73° + \sin 21°$

2 $\frac{1}{2}\cos 75° + \frac{1}{2}\cos 27°$

3 $\sin\frac{5\pi}{6} - \sin\frac{\pi}{6}$

4 $2\sin\frac{\pi}{12} - 2\sin\frac{7\pi}{12}$

5 $\frac{1}{2}\sin 145° - \frac{1}{2}\sin 13°$

6 $3\sin\frac{5\pi}{12} + 3\sin\frac{\pi}{12}$

7 $\frac{1}{4}\cos 4p - \frac{1}{4}\cos 10p$

8 $\cos 2A + \cos 2B$

9 $\sin 2P + 1$

10 $\frac{1}{2}\sin 6\theta$

11 $\cos 2\theta - 1$

Sums and differences to products (pp. 83–90)

1 Writing sums and differences as products (pp. 84–87)

1 $2\sin\frac{A+B}{2}\cos\frac{A-B}{2}$

2 $2\cos 46°\cos 23°$

3 $2\cos\frac{3\pi}{8}\sin\frac{\pi}{8}$

4 $-4\sin\frac{7\pi}{24}\sin\frac{\pi}{24}$

5 $2\cos\theta\cos 60°$

6 $2\sin 2\theta\cos\frac{\pi}{4}$

7 $2\sin\theta\sin 40°$

8 $2\cos 45°\sin(\theta - 45°)$ or $-2\sin 45°\sin(45° - \theta)$

9 $-4\sin 45°\sin(A - 45°)$ or $4\cos 45°\sin(45° - A)$

10 $2\sin\left(\frac{\pi}{4} + \frac{\theta}{2}\right)\cos\left(\frac{\pi}{4} - \frac{\theta}{2}\right)$

11 $\cos\left(\theta + \frac{5\pi}{12}\right)\sin\left(\theta - \frac{\pi}{12}\right)$

12 $6\cos^2\left(\frac{\theta}{2} + \frac{\pi}{2}\right)$

2 Proofs (pp. 88–90)

1 $\dfrac{\sin 80° - \sin 20°}{\cos 80° + \cos 20°} = \dfrac{2\cos 50°\sin 30°}{2\cos 50°\cos 30°}$

$= \tan 30°$

2 $\dfrac{\cos\frac{2\pi}{3} - \cos\frac{\pi}{2}}{\sin\frac{2\pi}{3} + \sin\frac{\pi}{2}} = \dfrac{-2\sin\frac{7\pi}{12}\sin\frac{\pi}{12}}{2\sin\frac{7\pi}{12}\cos\frac{\pi}{12}}$

$= -\tan\frac{\pi}{12}$

3 $\sin\frac{\pi}{2} + \cos\frac{\pi}{3} = \sin\frac{\pi}{2} + \sin\left(\frac{\pi}{2} - \frac{\pi}{3}\right)$

$= 2\sin\frac{\pi}{3}\cos\frac{\pi}{6}$

ISBN: 9780170425728

4 $\cos 3\theta + \sin\theta - \cos\theta = (\cos 3\theta - \cos\theta) + \sin\theta$
$= -2\sin 2\theta \sin\theta + \sin\theta$
$= \sin\theta(1 - 2\sin 2\theta)$

5 $\sin 5\theta + 2\sin 3\theta + \sin\theta = (\sin 5\theta + \sin 3\theta) + (\sin 3\theta + \sin\theta)$
$= 2\sin 4\theta\cos\theta + 2\sin 2\theta\cos\theta$
$= 2\cos\theta(\sin 4\theta + \sin 2\theta)$
$= 2\cos\theta(2\sin 3\theta\cos\theta)$
$= 4\cos^2\theta\sin 3\theta$

6 $10\sin 4p\cos p - 10\cos 3p\sin 2p = 5\sin 3p + 5\sin p$
$= 10\sin 2p\cos p$

7 $2\sin 13d\sin 9d - 2\sin 6d\sin 2d = -\cos 22d + \cos 8d$
$= \cos 8d - \cos 22d$
$= 2\sin 15d\sin 7d$

8 $2\sin 5y\cos 2y - 2\sin 9y\cos 6y = \sin 7y - \sin 15y$
$= -(\sin 15y - \sin 7y)$
$= -2\cos 11y\sin 4y$

9 $\cos 7y\cos 2y - \sin 11y\sin 6y = \frac{1}{2}\cos 9y + \frac{1}{2}\cos 17y$
$= \frac{1}{2}\cos 17y + \frac{1}{2}\cos 9y$
$= \cos 13y\cos 4y$

10 $2\sin^2 80° - 2\sin^2 20° = 2\sin 80°\sin 80° - 2\sin 20°\sin 20°$
$= (\cos 0° - \cos 160°) - (\cos 0° - \cos 40°)$
$= \cos 40° - \cos 160°$
$= 2\sin 100°\sin 60°$

11 $\cos^2\frac{2\pi}{3} - \sin^2\frac{\pi}{2} = \frac{1}{2}\left(\cos\frac{4\pi}{3} + \cos 0\right) - \frac{1}{2}(\cos 0 - \cos\pi)$
$= \frac{1}{2}\left(\cos\frac{4\pi}{3} + \cos\pi\right)$
$= \cos\frac{7\pi}{6}\cos\frac{\pi}{6}$

Trigonometric equations (pp. 91–111)

Basic equations (pp. 91–92)

1 $\theta = 360n \pm 65°$
$\theta = -295°, -65°, 65°, 295°$

2 $\theta = 180n + (-1)^n \times 80°$
$\theta = 80°, 100°, 440°, 460°$

3 $\theta = 180n + 28°$
$\theta = -152°, 28°, 208°$

4 $\theta = n\pi + (-1)^n \times 0.58$
$\theta = 0.58, 2.5616$

5 $\theta = 2n\pi \pm 0.72$
$\theta = 0.72, 5.563, 7.003, 11.846$

6 $\theta = n\pi + 0.334$
$\theta = -9.091, -5.949, -2.808, 0.334$

7 $\theta = 2n\pi \pm \frac{\pi}{4}$
$\theta = -\frac{\pi}{4}, \frac{\pi}{4}, \frac{7\pi}{4}$

8 $\theta = n\pi - \frac{\pi}{3}$ or $n\pi + \frac{2\pi}{3}$
$\theta = \frac{2\pi}{3}, \frac{5\pi}{3}, \frac{8\pi}{3}, \frac{11\pi}{3}$

9 $\theta = n\pi - (-1)^n \times \frac{\pi}{6}$
$\theta = -\frac{5\pi}{6}, -\frac{\pi}{6}, \frac{7\pi}{6}, \frac{11\pi}{6}$

10 $\theta = 2n\pi \pm 0.25$
$\theta = -0.25, 0.25, 6.0332, 6.5332$

11 $\theta = n\pi + 0.48$
$\theta = 3.6216, 6.7632$

12 $\theta = 180n + (-1)^n \times 63°$
$\theta = 117°$

13 $\theta = n\pi + \frac{\pi}{6}$
$\theta = \frac{\pi}{6}, \frac{7\pi}{6}, \frac{13\pi}{6}$

14 $\theta = n\pi - 45°$
$\theta = 135°, 315°, 495°, 675°$

Rearrangements of general solutions (pp. 98–100)

1 $\theta = 45n + 13°$
$\theta = -77°, -32°, 13°, 58°$

2 $\theta = 180n \pm 44°$
$\theta = 44°$

3 $\theta = 22.5n + (-1)^n \times 7°$
$\theta = -29.5°, 7°, 15.5°$

4 $\theta = \frac{n\pi}{4} + 0.05$
$\theta = -1.5208, -0.7354, 0.05, 0.8354$

5 $\theta = \frac{2n\pi}{3} \pm 0.12$
$\theta = 0.12, 1.9744, 2.2144$

6 $\theta = \frac{n\pi}{2} + 0.15$
$\theta = -2.9916, -1.4208, 0.15, 1.7208$

7 $\theta = n\pi \pm 0.285 - 0.15$
$\theta = 2.3566, 2.9266$

8 $\theta = \frac{n\pi}{3} + (-1)^n \times 0.1805 - 0.\dot{6}$
$\theta = 0.2, 1.6083, 2.2944$

Solving trigonometric equations by factorising (pp. 101–108)

a By factorising directly or using the Pythagorean identities

1 $0, \frac{\pi}{2}, \frac{3\pi}{2}, 2\pi$

2 $0°, 101.3°, 180°, 281.3°, 360°$

3 $0, 1.8235, 4.4597, 2\pi$

4 $\frac{\pi}{4}, \frac{3\pi}{4}, \frac{5\pi}{4}, \frac{7\pi}{4}$

5 $0, 1.2490, \pi, 4.3906, 2\pi$

6 $0, \pi, 2\pi, 3\pi, 4\pi$

7 $\frac{\pi}{2}, \frac{3\pi}{2}, \frac{5\pi}{2}, \frac{7\pi}{2}$

8 $0, 2\pi, 4\pi$

9 $1.4056, \frac{3\pi}{4}, 4.5472, \frac{7\pi}{4}$

10 $\frac{\pi}{4}, 1.2490, \frac{5\pi}{4}, 4.3906$

b Using the double angle and other identities

1 $\frac{\pi}{3}, \pi, \frac{5\pi}{3}$

2 $\frac{\pi}{4}, \frac{\pi}{2}, \frac{5\pi}{4}, \frac{3\pi}{2}$

3 $0, 2.6780, \pi, 5.8195, 2\pi$

4 $0.5236, \frac{\pi}{2}, 2.6180, 3.6652, \frac{3\pi}{2}, 5.7596$

5 $0, 0.6155, 2.5261, \pi$

6 $\frac{\pi}{2}, 2.0344, 5.1761, \frac{3\pi}{2}$

c By converting sums and differences to products

1 $0, \frac{\pi}{2}, \pi, \frac{3\pi}{2}, 2\pi$

2 $0, \frac{\pi}{5}, \frac{\pi}{3}, \frac{2\pi}{5}, \frac{3\pi}{5}, \frac{2\pi}{3}, \frac{4\pi}{5}, \pi$

3 $\frac{\pi}{4}, \frac{5\pi}{4}$

4 $\frac{\pi}{12}, \frac{5\pi}{12}$

5 $\frac{\pi}{4}, \frac{3\pi}{4}, \frac{5\pi}{4}, \frac{7\pi}{4}$

6 $0, \frac{\pi}{3}, \frac{2\pi}{3}, \pi$

d Solving equations by squaring

1 1.208, 5.719, 7.49

2 0.3070, 4.0106, 6.5902

Putting it together (pp. 112–119)

1 a Horizontal axis = d = $\frac{\text{max} + \text{min}}{2} = \frac{17 + 7}{2} = 12$

Graph is descending from y-axis, so a is negative, and amplitude = a = $\frac{\text{max} - \text{min}}{2} = \frac{17 - 7}{2} = 5$

Horizontal shift = c = 0 because the graph is at a minimum at the y-axis.

Frequency = b = $\frac{2\pi}{\text{period}} = \frac{2\pi}{2} = \pi$

$\therefore$ Equation is $h = -5\cos \pi t + 12$

b $12 - 5\cos \pi t = 9$

$5\cos \pi t = 3$

$\cos \pi t = 0.6$

$\alpha = 0.9273$

$\pi t = 2n\pi \pm 0.9273$

$t = 2n \pm 0.2952$

c After 0.2952 s and 1.7048 s

2 a Amplitude = 5

Period = $\frac{2\pi}{b} = \frac{2\pi}{1} = 2\pi$

Lateral shift = 0.6435 to the right

Vertical shift = 0

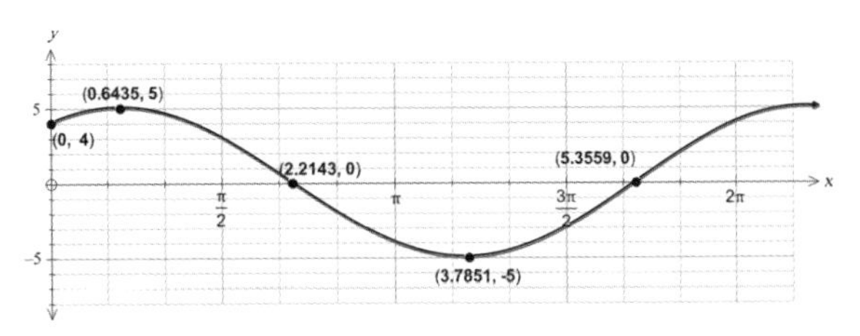

Maximum: (0.6435, 5)

Minimum: $(\pi + 0.6435, -5) = (3.7851, -5)$

y intercept: (0, 4)

x intercepts:

$5\cos (x - 0.6435) = 0$

$\cos (x - 0.6435) = 0$

$\alpha = 1.5708$

$x - 0.6435 = 2n\pi \pm 1.5708$

$x = 2n\pi \pm 1.5708 + 0.6435$

$n = 0 \Rightarrow x = 2.2143$

$n = 1 \Rightarrow x = 5.3559$

$\therefore$ x intercepts are (2.2143, 0) and (5.3559, 0)

b $4\cos x + 3\sin x = 0$

$3\sin x = -4\cos x$ (÷ by cos x)

$3\tan x = -4$

$\tan x = -1.3$

$\alpha = -0.9273$

$x = n\pi - 0.9273$

$n = 1 \Rightarrow x = 2.2143$

$n = 2 \Rightarrow x = 5.3559$

c $y = 5\cos (x - 0.6435)$

$= 5(\cos x \cos 0.6435 + \sin x \sin 0.6435)$

$= 5(0.8\cos x + 0.6\sin x)$

$= 4\cos x + 3\sin x$

d $x = 2n\pi \pm 1.1593 + 0.6435$

$x = 1.8028$ or 5.7674

e $4\cos x + 3\sin x = 2$

$4\cos x = 2 - 3\sin x$

$16\cos^2 x = 4 - 12\sin x + 9\sin^2 x$

$16 - 16\sin^2 x = 4 - 12\sin x + 9\sin^2 x$

$(\cos^2 x = 1 - \sin^2 x)$

$25\sin^2 x - 12\sin x - 12 = 0$

$\therefore$ $x = 0.9732$ or $x = -0.4932$

$\alpha = 1.3388$ $\qquad \alpha = -0.5158$

$x = n\pi + (-1)^n \times 1.3388$ $\qquad x = n\pi + (-1)^n \times -0.5158$

$n = 0 \Rightarrow x = 1.3388$

$n = 1 \Rightarrow x = 1.8028$ $\qquad x = 0.5158$

$n = 2 \Rightarrow x = 4.9444$ $\qquad x = 5.7674$

These answers need to be checked in the original equation because squaring an equation can give rise to extra incorrect answers.

$4\cos 1.3388 + 3\sin 1.3388 \neq 2$ ✗

$4\cos 1.8028 + 3\sin 1.8028 = 2$ ✓

$4\cos 0.5158 + 3\sin 0.5158 \neq 2$ ✗

$4\cos 4.9444 + 3\sin 4.9444 \neq 2$ ✗

$4\cos 5.7674 + 3\sin 5.7674 = 2$ ✓

$\therefore$ The only correct answers are $x = 1.8028$ and $x = 5.7674$ (same answers as in **d**).

ISBN: 9780170425728

3 a Amplitude = a = $\dfrac{15.75 - 8.65}{2} = 3.55$

b = frequency = $\dfrac{2\pi}{365}$

Lateral shift = left 9 ⇒ c = +9

Vertical shift = $\dfrac{15.75 + 8.65}{2} = 12.2$

Equation: $y = 3.55 \cos \dfrac{2\pi}{365}(x + 9) + 12.2$

b $x = 365n \pm 60.36 - 9$
Day 51 and day 295

c Equation: $y = 2.53 \cos \dfrac{2\pi}{365}(x - 9) + 12.13$

d Days 86 and 261. On both days they have 12 hours daylight.

4 a

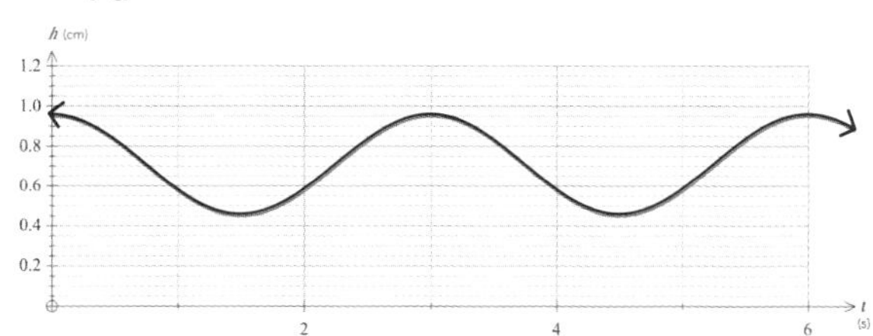

b 3 seconds

c $0.7 - 0.25\sin\left(\dfrac{2\pi}{3}(t - 0.75)\right) = 0.6$

$\sin\left(\dfrac{2\pi}{3}(t - 0.75)\right) = 0.4$

$\alpha = 0.4115$

$\dfrac{2\pi}{3}(t - 0.75) = n\pi + (-1)^n \times 0.4115$

$t = \dfrac{3n}{2} + (-1)^n \times 0.1965 + 0.75$

∴ Height is below 0.6 m between 0.9465 and 2.0535 seconds and between 3.9465 and 5.0535 seconds.

d $h = 0.7 - 0.25\sin \pi(t - 0.75)$

e $h = 0.3 - 0.2\sin\left(\dfrac{4\pi}{3}(t - 0.375)\right)$

Practice tasks (pp. 120–125)

Practice task one (pp. 120–121)

Model for curve A:

Using the model $y = a\cos bx \pm d$:

Amplitude of 2 ⇒ model takes the form $y = \pm 2\sin bx \pm d$

Frequency = $\dfrac{2\pi}{2\pi}$ ⇒ model takes the form $y = \pm 2\sin 1x \pm d$

Curve is decreasing immediately to the right of the origin ⇒ $y = -2\sin 1x \pm d$

∴ $\mathbf{y = -2\sin x + 0.5}$

Model for curve B:

Maximum where $y = 1$ and minimum where $y = -1$ ⇒

Amplitude = 1 and vertical shift = 0, so $y = 1\cos b(x \pm c) + 0$

Minimum where $x = 0$ and maximum where $x = \dfrac{\pi}{2}$

⇒ period = $2 \times \dfrac{\pi}{2} = \pi$, so the frequency = $\dfrac{2\pi}{2} = 2$

∴ $y = 1\cos 2(x \pm c) + 0$

Minimum at (0, –1) → no lateral shift so

$y = 1\cos 2(x \pm 0) + 0$

Curve is increasing immediately to the right of the origin ⇒ $y = -1\cos 2(x\, 0) + 0$

∴ $\mathbf{y = -\cos 2x}$

Particular solutions for the intercepts of $y = -2\sin x + 0.5$ and $y = -\cos 2x$ for $x < 4\pi$:

(0.5236, –0.5), (2.6180, –0.5), (6.8068, –0.5) and (8.9012, –0.5)

∴ Shading occurs between the points (0, –0.5) and (0.5236, –0.5), between (2.6180, –0.5) and (6.8068, –0.5), and between (8.9012, –0.5) and (4π, –0.5).

General solution for the intercepts of $y = -2\sin x + 0.5$ and $y = -\cos 2x$:

$-\cos 2x = -2\sin x + 0.5$

$2\sin^2 x - 1 = -2\sin x + 0.5$

$2\sin^2 x + 2\sin x - 1.5 = 0$

∴ $\sin x = 0.5$

or $\sin x = -1.5$, which is not possible.

$\alpha = 0.5236$

$\mathbf{x = n\pi + (-1)^n \times 0.5236}$

Practice task two (pp. 122–123)

Graphical features:

Using the model $y = a\cos b(x \pm c) \pm d$:

Maximum at (0, 1) and minimum at (0.7854, –3) ⇒

Amplitude = a = 2, and vertical shift = d = –1.

Horizontal difference between the maximum and minimum = 0.7854 ⇒ frequency = b = $\dfrac{2\pi}{2 \times 0.7854} = 4$

Maximum at (0, 1) ⇒ no horizontal shift, so c = 0

Putting these together produces the equation $y = 2\cos 4x - 1$, which shows that Fred is correct.

Comparing graphical features for one cycle ($0 < \theta < \dfrac{\pi}{2}$):

***y* intercepts:**

For $y = \cos 4\theta - 2\sin^2 2\theta$: $y = \cos 4 \times 0 - 2\sin^2 0 = 1$

For $y = 2\cos 4\theta - 1$: $y = 2\cos 4 \times 0 - 1 = 1$

∴ *y* intercepts are the same.

x intercepts:

For $y = \cos 4\theta - 2\sin^2 2\theta$:

$\cos 4\theta - 2\sin^2 2\theta = 0$

$2\sin^2 2\theta = \cos 4\theta$

$2\sin^2 2\theta = 1 - 2\sin^2 2\theta$ (Using double angle identity)

$4\sin^2 2\theta = 1$

$\sin^2 2\theta = 0.25$

$\sin 2\theta = 0.5$

$\alpha = 0.5236 = \frac{\pi}{6}$

$\therefore\ 2\theta = n\pi + (-1)^n \times \frac{\pi}{6}$

$\theta = \frac{n\pi}{2} + (-1)^n \times \frac{\pi}{12}$

$n = 0 \Rightarrow \theta = \frac{\pi}{12}$

$n = 1 \Rightarrow \theta = \frac{\pi}{2} - \frac{\pi}{12} = \frac{5\pi}{12}$

For $y = 2\cos 4\theta - 1$:

$2\cos 4\theta = 1$

$\cos 4\theta = 0.5$

$\alpha = 1.0472 = \frac{\pi}{3}$

$\therefore\ 4\theta = 2n\pi \pm \frac{\pi}{3}$

$\theta = \frac{n\pi}{2} \pm \frac{\pi}{12}$

$n = 0 \Rightarrow \theta = \frac{\pi}{12}$

$n = 1 \Rightarrow \theta = \frac{\pi}{2} - \frac{\pi}{12} = \frac{5\pi}{12}$

$\therefore$ x intercepts for both equations are $\left(\frac{\pi}{12}, 0\right)$ and $\left(\frac{5\pi}{12}, 0\right)$ which are the same.

The x and y intercepts are the same, and we know the maxima and minima are the same, so these also show that Fred was correct.

Proof that $y = \cos 4\theta - 2\sin^2 2\theta$ and $y = 2\cos 4\theta - 1$ are the same:

$\cos 4\theta - 2\sin^2 2\theta = \cos 4\theta - (2 \times \sin\theta \times \sin\theta)$

$= \cos 4\theta - (\cos 0 - \cos 4\theta)$

(double angle identity)

$= \cos 4\theta - 1 + \cos 4\theta$

$= 2\cos 4\theta - 1$

Practice task three (pp. 124–125)

Graph:

Axis of symmetry = vertical shift = d = 15.8

Maximum and minimum: c = lateral shift: = +1.65 ⇒ sine curve moves left 1.65 months ⇒ maximum is at a quarter of the cycle (3) – 1.65 = 1.35. Minimum is at 6 + 1.35 = 7.35.

Temperature at the maximum = 15.8 + amplitude = 15.8 + 4.3 = 20.1

Temperature at the minimum = 15.8 – amplitude = 15.8 – 4.3 = 11.5

$\therefore$ Maximum at (1.35, 20.1) and minimum at (7.35, 11.5)

y intercept: $m = 0 \Rightarrow T = 4.3\sin\frac{\pi}{6} \times 1.65 = 19.07$

Frequency: $\frac{2\pi}{\text{period}} \Rightarrow \text{frequency} = \frac{\pi}{6}$

Points where graph crosses axis of symmetry: period = 12, so x coordinates are

1.35 + 3 = 4.35 and 1.35 + 9 = 10.35 ⇒ (4.35, 15.8) and (10.35, 15.8)

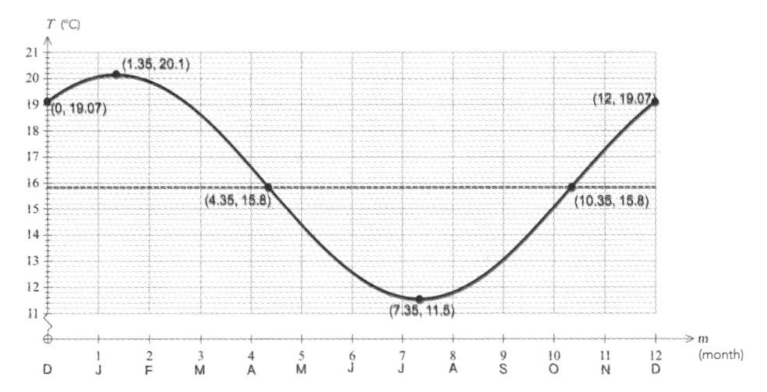

Periods with poor fit:

Substituting m = 0, m = 1, …, m = 11 into equation:

Month	Jan	Feb	Mar	Apr	May	Jun	Jul	Aug	Sep	Oct	Nov	Dec
Temp (°C)	19.9	20.2	18.7	16.6	14.4	12.4	11.6	11.9	13.3	14.6	16.4	18.5
Model	20.0	19.9	18.6	16.6	14.4	12.5	11.6	11.8	13.0	15.0	17.2	19.1
Difference	–0.1	0.3	0.1	0	0	–0.1	0	0.1	0.3	**–0.4**	**–0.8**	**–0.6**

This shows that the model is a relatively poor fit in October, November and December, when the actual temperatures are between 0.4 and 0.8 lower than the model.

Estimate of earliest date to plant seed:

$16 = 15.8 + 4.3\sin\frac{\pi}{6}(m + 1.65)$

$\sin\frac{\pi}{6}(m + 1.65) = 0.0465$

$\alpha = 0.0465$ (unlikely but true)

$\frac{\pi}{6}(m + 1.65) = n\pi + (-1)^n \times 0.0465$

$m = 6n + (-1)^n \times 0.0888 - 1.65$

$m = 4.2611$ or 10.4388

$\therefore$ The model estimate for the earliest he should plant is at month 10.4388, which is about the fourteenth of October. However, between October and the end of the year, the model was about half a degree above the actual mean temperatures. Therefore the temperature is unlikely to be at 16° on that date, and the farmer would be wise to plant at least a week after that date.

ISBN: 9780170425728